Jörg Sander

Coil Coating

Cover: Salzgitter Stahl, Salzgitter/Germany

Jörg Sander
Coil Coating
Hanover: Vincentz Network, 2014
European Coatings Tech Files
ISBN 978-3-86630-834-3

Please ask for our book catalogue
Vincentz Network, Plathnerstr. 4c, 30175 Hanover, Germany
T (202) 684-6630, F (202) 380-9129
E-mail: books@american-coatings.com, www.american-coatings.com

Layout: Danielsen Mediendesign, Hanover, Germany

ISBN 978-3-86630-834-3

EUROPEAN COATINGS TECH FILES

Jörg Sander

Coil Coating

Foreword

During my entire business career, I have seen many industrial applications of paints and coatings to metal, be it on extruded profiles, bicycle frames, escalator steps, car bodies, wheel rims and automotive trim parts, or beverage cans, to name a few. The most appealing application for me, however, has always been the continuous coating of metal coils.

Coil coating provides fascinating ways of producing coated metal goods. Highly sophisticated processes and the proficient and sustainable use of elaborate process chemicals and materials bring about durable, high-quality painted products. These can be processed to create a wide range of articles with an abundance of uses, technical features, shapes, and surface aspects.

This present book is based on the manuscript of a tutorial I held during the European Coatings Conference Coil and Can Coating that was organised by Vincentz Network in Berlin in October 2011. It is the first comprehensive English-language description of the coil coating technology available to the public in years. To make the book attractive for the insiders of the coil coating community and instructive for newcomers to this exciting technology alike, a good deal of elaboration was needed beyond the conference script. Thus the book embraces both: A synopsis, maybe also a useful look at the bigger picture for those already conversant with the subject, and an in-depth introduction for those who want to familiarise with it. In particular, students of process and equipment engineering, surface science, pretreatment and paint technologies, but also of architecture and metal design shall find detailed, and, I hope, popular descriptions, explanations, and interesting connections demonstrated.

I am grateful to many colleagues in the business who helped in discussions and by providing detail information about their specialist subjects. A particular pleasure was meeting Dr. Bernd Meuthen, Past President of the ECCA, long-standing President and Honorary President of the German ECCA Group, and lead author of a German standard publication on the topic, during my very early days in the business. I have been very much indebted to Bernd for sharing his profound knowledge and his enthusiasm for coil coating that I have always admired. The compilation on standards, one of his very particular interests, is based on his earlier work. Sadly, Bernd Meuthen died on October 24, 2013. And so I feel it fitting to dedicate this work to his memory. I hope it helps to inspire and enthuse future coil coating chemists and engineers as Bernd was able to do.

Velbert, Germany, October 2013

Jörg Sander

Contents

1 Coil coating - principle, market and applications

1.1 Introduction

Coil coating makes use of a simple, but effective principle, i.e. to clean, pretreat and coat flat coils or sheets of steel or aluminium in a continuous operation, before other stages of industrial manufacture [1-4]. The slogan for this concept is: Finish first - fabricate later!

A conventional process sequence in coil coating consists of:

- Cleaning
- Conversion treatment
  (including optional post-rinse)
- Drying
- Primer coating
- Top coating
- Foil lamination (optional)

An overview of a large coil coating line is depicted in Figure 1.1.

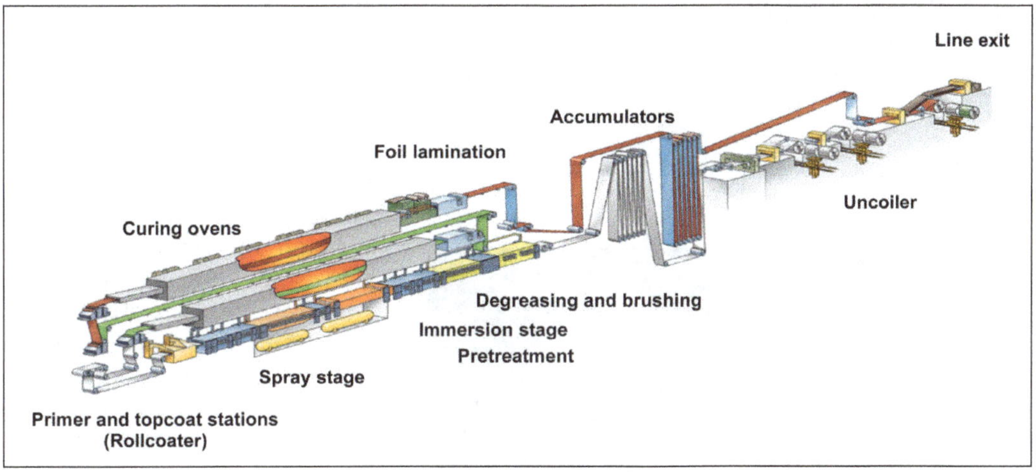

Figure 1.1: Schematic of a steel coil coating line Source: Voestalpine

The official definition of the coil coating process is codified in the European Standard EN 10169-1: 2003. It states that coil coating is

- "a method in which an organic coating material is applied on rolled metal strip in a continuous process. This process includes cleaning if necessary and chemical pre-treatment of the metal surface and either one side or two side, single or multiple application of (liquid) paints or coating powders which are subsequently cured or laminated with plastic films".

1.2 Coil coating process

Metal coils are fed into a coil coating line from an entry station, where the coils are reeled off an uncoiler (Figure 1.2), and the beginning of one coil fixed to the end of the preceding coil by welding or stitching. The uncoiler station is usually duplicated, or equipped with twin reels, in order to allow for faster coil changes. The metal strip then enters the active section of the line through a loop accumulator. The large accumulator shown in Figure 1.3a has been emptied during the change of a coil, while feeding the strip length that was stored in it to the continuously running process. It waits to be filled up again to be ready for the next coil change.

Figure 1.2: Coil reel Source: ThyssenKrupp Steel Europe

Figure 1.3: a) Entry accumulator; b) Loop pit

Source: a) ThyssenKrupp Steel Europe; b) Alcoa Inc.

The active section comprises a degreasing stage, with a brushing section and rinses; one or more pretreatment stages in either immersion or spray technique, followed by another rinse cascade, and a water dryer. In case of so-called no-rinse pretreatment processes, the strip needs to be dried beforehand, and the treatment solution is applied by a roll-coater (chem-coater), before reacting off and evaporating the water in the dryer. The pretreated strip then enters the primer stage with its roll-coater unit and curing oven, followed by a water quench cooler. Afterwards, the finish coater section is passed which again comprises the roll-coater units and curing oven with subsequent cooling. The finish (topcoat) section often has two coater installations that allow quicker colour changes. After the topcoat, a lamination with protective foil may follow, before the strip leaves the active section via the exit accumulator. Finally, the strip passes an inspection stand and is taken off the line in an exit unit with a shear and a recoiler.

Figure 1.4: No-rinse chem-coater in a pretreatment stage

Source: Salzgitter Stahl

The entry and exit accumulators are operated to allow a uniform speed throughout the active stages. They are usually dimensioned to provide material for 1 to 2 min of production time stored in them.

Smaller and slower coil coating lines may be equipped with a simpler installation, like the loop pit in an aluminium strip line shown in Figure 1.3b. Similar pits may also be used as an inspection stand.

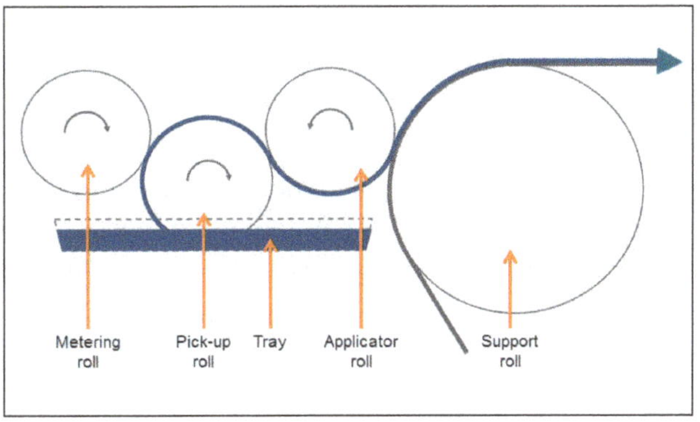

Figure 1.5: Schematic of a three-roller paint coater

In high-capacity lines, coils are handled in sizes up to over 2 m width and 2 m diameter. Dimensions in steel lines are usually smaller due to the 3-fold density and weight of steel as compared with aluminium. Still, such a coil weighs up to 20 tonnes.

Figure 1.4 illustrates a roll-coater machine which is used to apply a no-rinse pretreatment. As it does not process any solventborne paint, it needs not be enclosed in a separate compartment that would be necessary to keep the flammable solvent vapours from the working environment. Roll-coaters for pretreatment are also called chemcoaters, as they deal with low-viscosity, aqueous chemical solutions and are adapted to this purpose.

Any roll-coater is designed to pick up an amount of liquid, and transfer it to the moving strip. Usually, there are two- or three-roller arrays. Three-roller coaters may have the rolls mounted in V-shape or with their axes in a line. The nip pressure between the rollers, their relative rotating speed, and their direction of rotation are manipulated in order to obtain a uniform wet film of the desired thickness across and along the strip.

Figure 1.5 shows the schematic of a V-shaped three-roller device in forward operation. This means that the applicator roll is rotated so that its surface moves the same direction as the strip, with only little difference of the surface speed. Liquids with lower viscosities, including aqueous pretreatment solutions, require the applicator to be operated in reverse mode. Since the applicator roll is clad with a rubber or elastomer blanket of adjusted hardness, reverse operation leads to high wear of the roller surface because of the strip edges cutting into it.

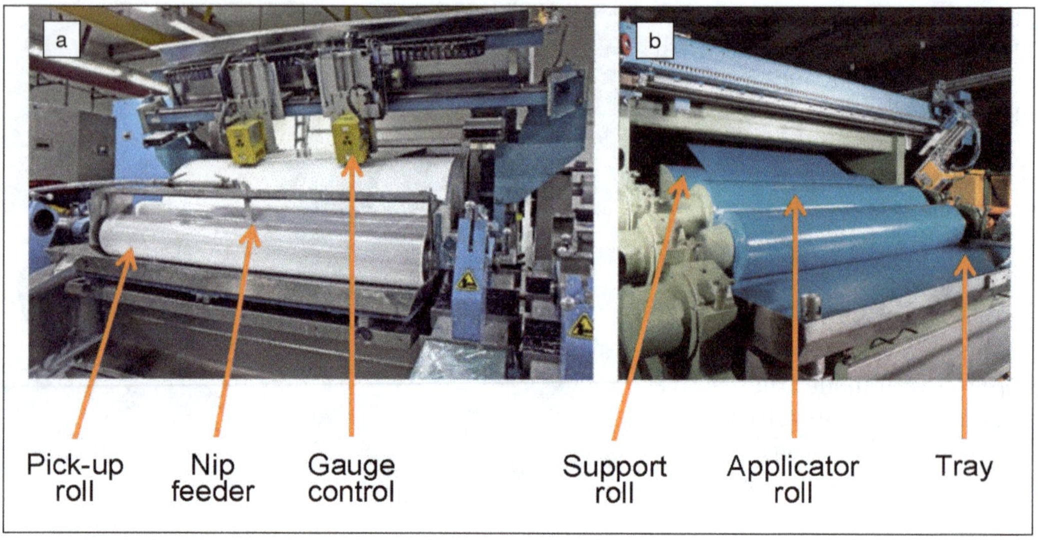

Figures 1.6a and b: Top-side paint roll-coaters in operation Source: a) Salzgitter Stahl, b) ThyssenKrupp Steel Europe

The pick-up roll, usually made with a ground stainless steel or hard chromium-plated surface, takes the liquid from a reservoir tray, the wet film is reduced by the metering roll (or, alternatively, a fixed knife, the so-called „doctor blade"), then transferred via the applicator roll onto the strip.

Roll-coaters for the top side of the strip are depicted in Figure 1.6. While on the left (Figure 1.6a), the liquid is fed directly into the nip between the two rolls, the right coater (Figure 1.6b) is operated with a tray from where the paint is taken.

Mounted either above the applicator roll or the support roll, the heads of the gauge control equipment can be seen. They are either fixed to a position across the strip, or they can be moved across statistically or in regular oscillation, to obtain full-area continuous monitoring of the wet film. The measuring principle can be based on infra-red reflection, or electron backscattering induced by a weak radioactive source, krypton-85 (^{85}Kr).

The wet, coated strip enters the curing oven immediately after the coater house. Most ovens are operated on hot-air convection with air temperatures up to 400 °C. Some lines are equipped with IR, near-IR or induction ovens.

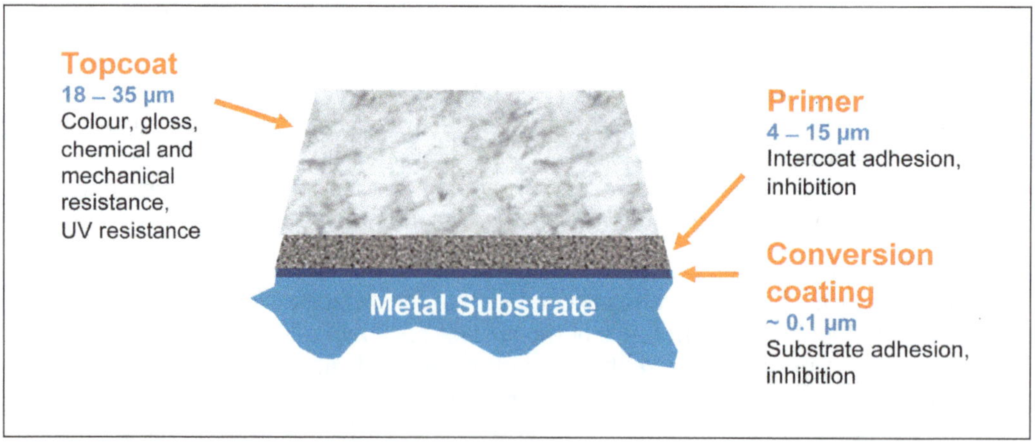

Figure 1.7: Typical build-up of a coil paint coating

Liquid paints in coil coating are mostly applied in two layers, i.e. a primer and a topcoat layer. The majority of primers are applied at 5 µm dry film thickness. A normal topcoat system has a gauge of approx. 20 µm (cf. Figure 1.7). For high endurance requirements, however, both primer and topcoat may be used at higher gauges.

There are also single-coat systems available (particularly for uses on Al substrates, and multiple coats may be required for specialist purposes. Backing coats are either applied on a regular primer, or as single coats at typical thicknesses of 10 to 15 µm dry film.

Figure 1.8: Removal of a finished coil from the line

Source: Salzgitter Stahl

The finished coated coils are taken off the recoiler at the line exit, packed and labelled, and stored prior to transport to the end user, as illustrated in Figures 1.8 and 1.9.

Protective foils are applied optionally as the last step of the coil coating sequence, in order to mechanically protect the finished surface from damages. They are stripped, after the metalworking process, from the end product.

In the case of decorative foils used as the finish coat, these are applied on top of a primer or co-laminated onto an intermediate coating.

Figure 1.9: Finished coil storage

Source: Salzgitter Stahl

PVC films are used for decorative purposes providing particular surface patterns like woodgrain, leather aspect, stone imitations, etc.

Polyethylene terephthalate, PET, or polyester films are hot-laminated onto the last paint layer. Being transparent or coloured and patterned themselves, they allow a multitude of attractive surface finishes, colours, imprints, and gloss grades, in addition to particular technical requirements like high flexibility, scratch resistance, resistance against aggressive environments, detergents, solvents, etc., anti-grafitti features, or foodsafe certification. They are also particularly capable of deep drawing without gloss reduction. Coil coated metal, finished and recoiled, is ready to be formed into the ultimate part by bending, rolling, drawing, punching, etc.

1.3 Metalworking with precoated coil

Prepainted metal can undergo numerous processing steps to create the final three-dimensional workpiece that is fabricated. Some examples are depicted in Figure 1.11. The surface being already that of a finished product, the prepainted stock should be handled with appropriate attention. When proper manufacturing techniques are used, prepainted metal is as easily transported, stored and handled as other material. The processing tools need to be adapted, and therefore should be dedicated to prepainted feedstock. For instance, tool surfaces must have smoothly polished and considerably hardened surfaces, in particular when they touch the visible faces of the final item. Feeding must allow traction without skidding, and continuous operation without stoppages, i.e. using a loop feeder with suitable speed regulation.

Usually, prepainted sheet does not require extra lubrication for forming. However, the clearances of the tools, like rollers, dies and cutters must be adjusted to the additional paint film thickness. For all deforming operations, working at temperatures above the glass transition temperature of the coating is advantageous. The forming capability of the prepainted metal can be assessed by standardised simulation methods.

In many cases, **roll-forming** is employed to form corrugated or grooved panels, profiles of any shape, or tubes.

Bending and flanging are common ways to fold up sheet ends or to join two panels together. As any cut edge of precoated metal is inadvertently left unprotected, bending and flanging is often used to retract these edges from the surrounding.

Deep-drawing must consider the material flow including the flexibility of the paint to ensure the deformation limits are not exceeded.

Forming corners from a precoated sheet is a particular design task. For small deformations, corners can be made by deep-drawing. In other cases they are obtained by folding operations. According to the design and quality requirements, the folding can be done to result in open, diagonal, or mitred corners. The latter involves both flanks of the bent sheet to be folded in so none of the cut edges are left open to view.

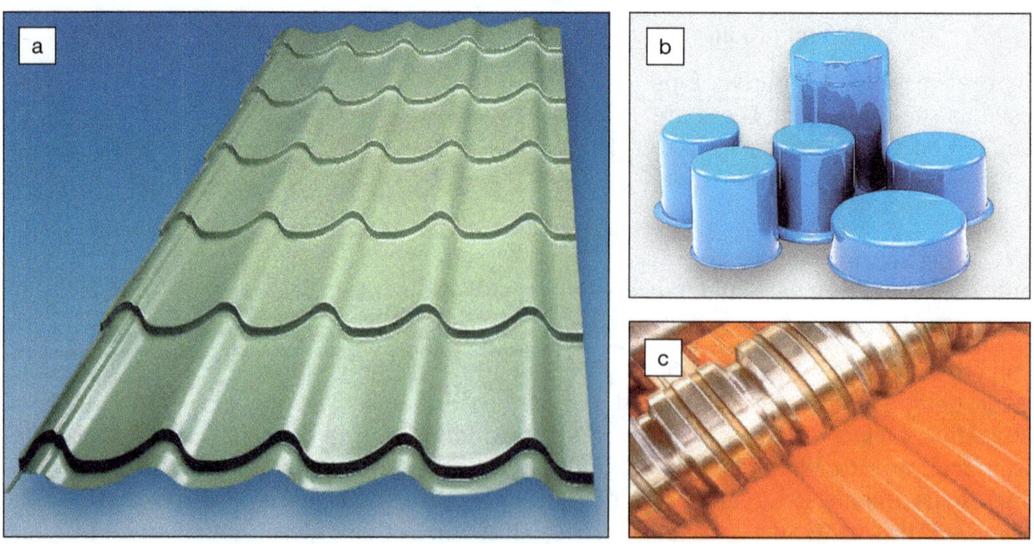

Figures 1.10a to c: Examples of roll-formed and deep-drawn goods made from precoated coil

Source: a, b) ThyssenKrupp Steel Europe, c) ArcelorMittal

Cutting and punching of precoated metal requires properly sharpened tools. If possible, the cut or punch should be performed from the coated side.

There are a variety of ways for fastening and **joining** of prepainted metal. Adhesive bonding is often chosen when it comes to combination with other materials. **Mechanical fastening** with screws, bolts or rivets is a more obvious method that however bears in it the disadvantage of drilling holes through the sheet material first. Integral joining by clinching or stitching provides a more elegant way. Clinching requires somewhat thicker sheets to be joined, as it involves the material of the overlapping sheets to flow and form the interlocking buckle and cavity. It can be performed so that the joint is invisible from the outer surface.

Table 1.1: Common forming and joining techniques for precoated coil

Source: ECCA Academy, 2011

Metalworking with precoated coil

- **Roll-forming**
  - Most common 3-D shaping method for coil and sheet metal
  - Smooth and hardened tool surfaces required
- **Bending, flanging**
- **Deep-drawing**
- **Forming corners**
  - Deep-drawing
  - Folding
    open corner (cut edges visible)
    diagonal corner (cut edge visible on one flap)
    mitred corner (no visible cut edges)
- **Punching and shearing**
  - Properly sharpened tools required
- **Joining**
  - Adhesive bonding (incl. joining with other materials)
  - Mechanical fastening (screws, bolts; rivets)
  - Integral joining techniques (clinching, stitching)
  - Lock-forming (standing seam, flanging)
  - Welding (resistance welding)

Panels are bent and flanged together to form fold-up seams that are either visible (like a standing seam) or retracted from sight.

Even resistance **welding** is possible, providing the coating does not insulate too much. It needs either applying the coating only on one side or using thin or special conductive coatings. With the proper conditions, welding can be performed on the reverse of a sheet without leaving traces on the visible outer face.

1.4 Inherent benefits of coil coating

Coil coating provides a lot of benefits when compared to post-finishing.

First of all, the **quality** of the paint finish is consistently high, because the process is run continuously and highly automated. All along and across the coated strip, the coating will have a uniform thickness and appearance within narrow tolerances. Gauge variations are in the range of <1 µm. Regarding durability in terms of adhesion, corrosion and weathering resistance, there are architectural products available with lifetime warranties up to 40 years. To avoid mechanical damages of the prefinished metal during onward processing, protective foils can be applied. These stay on the coil until after the metalworking and mounting and are stripped afterwards.

For the user of precoated coil, it is important that the design of end products is easy and flexible, once some principles for the processing of the prefinished feedstock are adopted. Choice can be made from an abundance of colours and finishes. Prepainted metal also protects areas that would go uncovered by post-finishing. Working with prefinished metal produces much less pollution and waste than a post-finishing concept.

Table 1.2: Benefits inherent to the coil coating process

Benefits in coil coating
• **Quality**
- Highly automated continuous process
- Consistent quality of paint finish
- Precise specifications (gauge, appearance, performance)
- Consistent high standard of adhesion/corrosion resistance
- Durability warranties up to 40 years service life (architecture)
- Protective film applicable to minimise handling damage
• **Processing**
- Product design easier and more flexible
- Hidden areas protected by prepainted metal
- Wide range and consistency of finish colours
- Minimal pollution with little waste products
• **Economy**
- Cost effective for the coil coater: Low use of consumable materials High coating transfer efficiency Custom width substrate
- Cost effective for the end user: no in-house pretreatment or paint processes, incl. supply media and waste lower capital investment requirements less factory space (operation and storage areas) less inventory reduced energy consumption, waste, effluent, emissions smaller labour force lower insurance costs (no storage of hazardous chemicals)
• **Recyclability**
- Fully recyclable, scrap infrastructure exists
• **Safety**
- Highly automated process intrinsically safe
• **Environment**
- Low and well controlled emissions, incl. waste treatment and disposal, effluent treatment, VOC, CO_2

Benefits also translate to the **economy and ecology** of the process.

Coil coatings are applied with an optimised consumption of materials, and to almost all customary sheet dimensions.

The user of prepainted stock enjoys multiple savings just because he does not need to operate the pretreatment and painting process. This means there is a lot less of consumables and waste, there is lower investment, floorspace, inventory and energy consumption. There are no concerns about waste and effluent treatment and disposal, and no emissions of volatiles. There is less, and safer, labour, because no hazardous or inflammable materials need to be handled, and finally, this also results in lower insurance cost.

Precoated sheet and coil is entirely recyclable, and there is a functioning infrastructure in place to collect and recycle metal scraps.

All in all, coil coating, as an automated, well-controlled and low-impact process, has a built-in safety, for the manufacturer, the operator, the user, as well as for environment.

1.5 Coil coating market in Europe

Looking at the coil coating market in Europe [5, 6], nowadays, 1.1 billion m^2 of mostly galvanised steel and 220 million m^2 of aluminium strip are precoated. These figures must be understood as the dimensions of the metal, while the actually coated surface is double, since top and reverse sides have to be considered. It should also be mentioned that, by definition, these figures do not enclose the production of coated steel and aluminium for food and beverage packaging uses, which in terms of the process technology are very similar.

The European market comprises some 200 coil coating lines, of which approx. 60 are high-capacity lines with tonnages of 200,000 t/a and over.

Fourteen steel lines are coupled with a galvanising installation, two Al lines have an annealing section upfront. Many of the smaller lines are small-strip installations that produce, for instance, packaging strip or venetian blinds. Currently, powder paints are used in thirteen coil coating operations; the vast majority of lines use liquid paints.

In Europe, most coil coating lines are owned by the large steel or aluminium producers who usually operate several of them in different locations. But there are also medium-sized enterprises.

Average dimensions in a steel coil coating line are a width of 1.6 m, and a metal thickness of 0.5 mm. A typical line speed is 100 m/min, with an upward trend. Dimensions for Al lines are typically larger, with widths of 2 m and speeds going up to 200 m/min.

Table 1.3: Key data of the coil coating market in Europe

Volume	1.1 billion m²/a steel (approx. 95 % galvanised) 220 million m²/a aluminium approx. 200 lines, hereof approx. 60 large lines (200 kt/a), 14 integrated galvanising, 13 powder
Operators	ArcelorMittal, Norsk Hydro, Tata Steel, ThyssenKrupp Steel Europe etc.
Ave. dimensions	approx. 1.60 m width, 0.5 mm thickness 100 m/min line speed
Paint chemistry	Polyester/melamine, polyurethane, PVdF, PVC
Market size	180 to 200 kt/a (approx. 900 million €/a) (topcoats, primers, backing coats, specialties)
Target markets	Architecture, transport, appliance, gen. industry

Major paint suppliers are the large groups that operate worldwide, however also here, some smaller companies participate in the market.

Overall, the potential paint usage is almost 200 kt/a, translating to a value of almost 1 billion €/a, including topcoats, primers, backing coats and some specialty coatings. The major target markets are architectural uses, transport and the domestic appliance industry. Precoated strip is also sold for a large variety of general industrial uses.

The graph (Figure 1.11) shows the historical development of shipments of precoated steel since 1970 in Europe. It should be mentioned that the tonnages of steel grew at a lower rate, because of the long-term trend to use thinner, stronger material.

While the recession of 2008 clearly took its toll, the market has largely recovered, but seems to stabilise, and not return to its former growth rates, 5 to 8 % year-on-year. The steel shipments have achieved a level of 1.1 billion m² again, although some of the recent growth might be attributed to stockbuilding. Over the past two decades, a considerable shift of demand to Central and Eastern Europe has taken place since the fall of the Iron Curtain in 1990, and it should be noted also that, since about 2005, imports from China, Korea, India and Taiwan have taken a share of approximately 15 % in the competition. Worldwide figures for the coil coating market are about 4 times the European data.

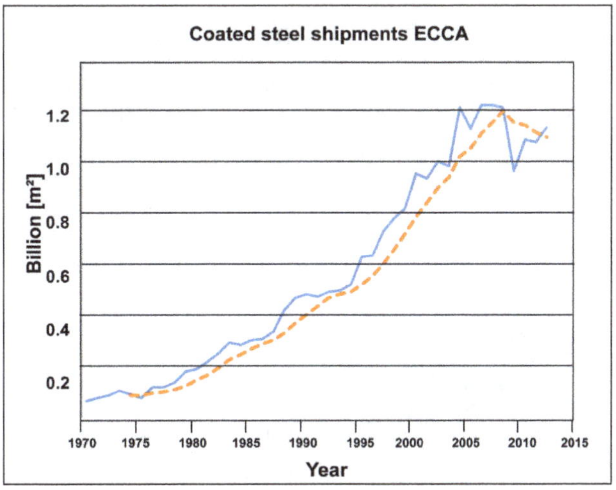

Figure 1.11: Historical development of shipments of precoated steel in Europe. Blue line: Annual shipments; orange line: Moving average Data source: ECCA Statistics 1970-2012

1.6 Applications of precoated metal

1.6.1 Overview

A closer look at the uses of precoated metal and the discussion of some examples will demonstrate the versatility of coil coating. As mentioned, the major target industries are the architectural sector with interior and exterior applications, the transport and the domestic appliances markets. General industry like the manufacture of metal furniture, climatisation equipment, teletronics and other domestic electrical gear also takes a considerable share.

Figure 1.12a shows the distribution of precoated steel uses per application. Architecture takes by far the largest share of over 70 %. A part of precoated material is distributed via service centres and stockists. The final destination and end-use for this material is therefore unknown, however it is safe to expect a similar distribution of this material as for the big rest.

An analogous view on precoated Al (Figure 1.12b) shows an even stronger segment dedicated to the buildings market. This includes production of roller shutters and sunblinds. The market for precoated aluminium semis is currently declining. In 2012, it suffered shrinkage of 8 to 9 % versus the previous year.

Despite the size of the coil coating market, and the undisputed benefits of the technology, there is still a large portion of potential end users left unpenetrated. Looking at the total market of galvanised steel coil that is produced in Europe (Figure 1.13), apparently only 20 % hereof actually are channelled through a coil coating process. Another 20 % are altogether left uncoated. 30 % are delivered to automotive OEMs with their specialist batch coating lines for car bodies, truck cabins and parts. The remainder of 30 %, however, is sold to manufacturers who produce single workpieces from coil or sheet blanks and have the painting of these goods either done inhouse or at job coaters.

A good part of this potential might indeed be penetrable for the coil coating technology. Reasons why coil coating did not yet enter may be manifold, from historical to processing to economic motivations. The latter often has to do with the economy of size resulting in supply restrictions and small-order surcharges for precoated metal.

1.6.2 Building applications

Uses in both indoor and outdoor architecture form the most important market. Precoated products are used for wall cladding and roofing of buildings. Almost traditional applications are large industrial halls and plant buildings, and agricultural buildings like stables, barns and sheds. With the grown acceptance and appreciation, nowadays also large administra-

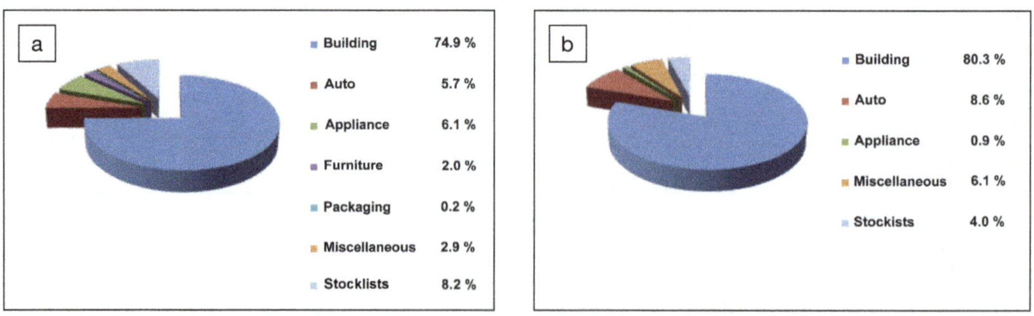

Figure 1.12: Uses of precoated metals in Europe; a) steel; b) aluminium Data source: ECCA Statistics 2012

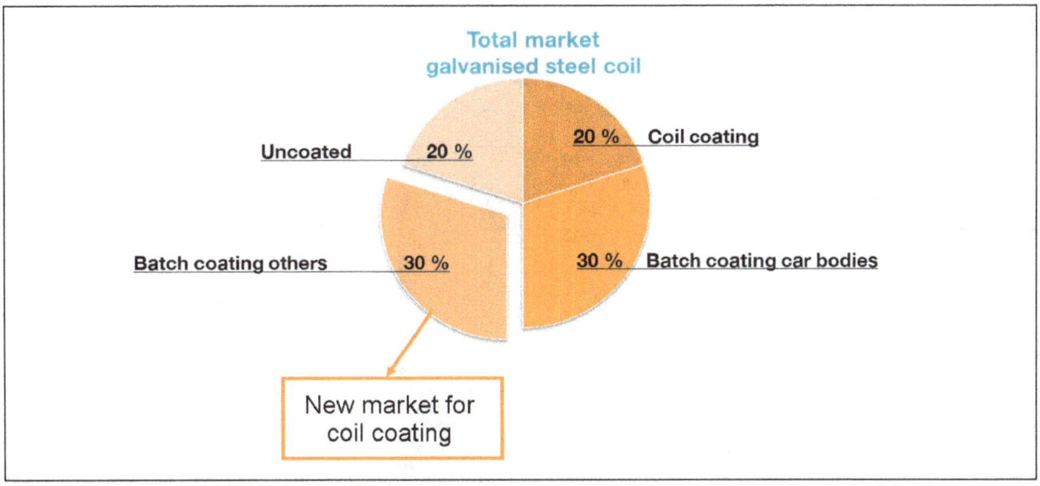

Figure 1.13: Potential expansion of precoated metal vs. batch-coating installations

tion and office buildings are made of precoated metal, their design taking advantage of the manifold possibilities in forms and surface aspects. Domestic building is a more recent market to be tackled. External applications also comprise garage doors, roller shutters and blinds, as well as composite panels with thermal or noise insulation.

Wall cladding is very commonly done with precoated metal on large industrial, agricultural or administration buildings, regardless whether the material is mounted in long strips of corrugated or V-groove panels, or used to fabricate single cassettes built into curtain walls. The photographs of Figure 1.14 show some examples.

Roofing with precoated metal in standing seam or tile fashion is popular in the domestic building market in some countries, while in others, the traditional ceramic or concrete tiles

Figure 1.14a to e: Applications of precoated metal in façade cladding of industrial and residential buildings

Source: a) ThyssenKrupp Steel Europe

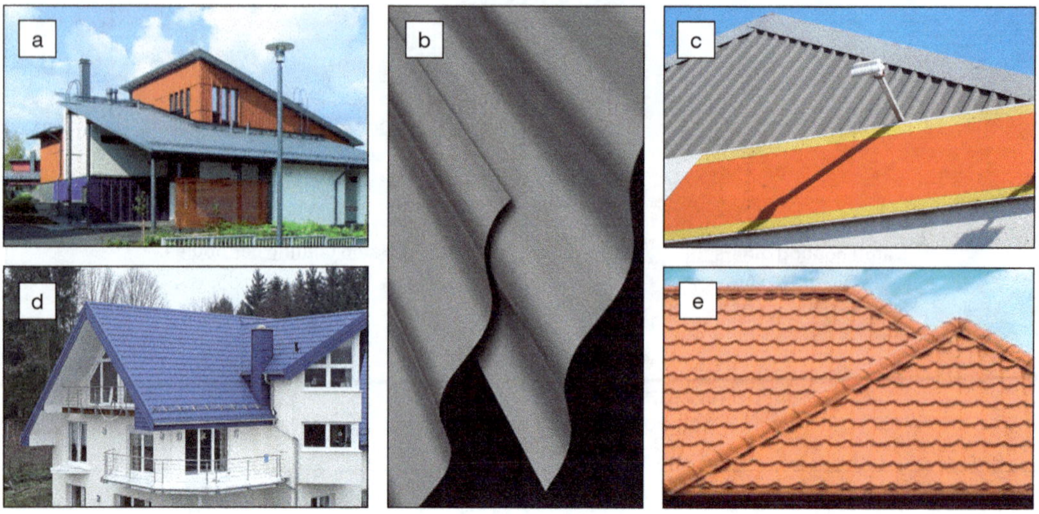

Figure 1.15a to e: Applications of precoated metal in roofing of residential buildings

Source: a) Ruukki; b,e) ArcelorMittal; d) ThyssenKrupp Steel Europe

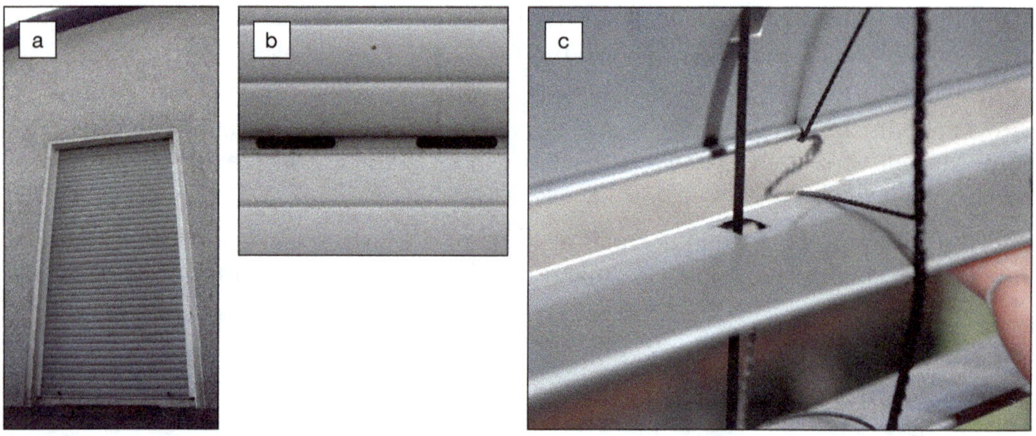

Figure 1.16a to c: Applications of precoated metal for shutters and blinds

Source: a,b) Sander, A.M.; c) Henkel

prevail. Figure 1.15 shows a variety of uses for roofing. Corrugated or standing seam appearance is often found on the traditional industrial or rural buildings, while in domestic housing, they imply an individual design touch. Tile-formed metal sheet is used in more conventional buildings, where it proves particularly convenient for roof repair systems.

To complete a metal-clad roof, metal flashings are obvious. Finally, precoated metal is also used for gutters, sinks and drains. Two other examples of outdoor building applications of precoated metal are shown in Figure 1.16. Roller shutters and sunshades traditionally are a domain of precoated aluminium.

Indoor applications include partitioning, wall panels, ceiling and flooring panels.

1.6.3 Transport applications

Precoated metal is also used in the **transport** sector. Typical products include sheet and panels for containers, caravan and mobile home sidings and interior, van and trailer bodies, insulated boxes, road and railway coaches (cf. Figure 1.17). Also small parts like license plates, wiper assemblies and trim are among the applications.

Figure 1.17a to c: Uses of precoated steel and aluminium in vehicle manufacture

Source: a) Euramax; b, c) ThyssenKrupp Steel Europe

A specialist use is the preprimed sheet that goes into passenger **car bodies**. A zinc-rich paint is applied on pretreated electro-galvanised steel. This coating, by its nature, provides improved corrosion resistance and compatibility with the processes in car body production that involve resistance welding and cathodic electro-dip painting. The enhanced corrosion protection is particularly important in critical areas of a car body, like seams, box sections or flanges, where a pretreatment is geometrically difficult to apply and may therefore be faulty, or the electro-dip paint is only insufficiently deposited due to shielding effects.

1.6.4 Applications in appliance and general industries

Another important use of precoated metal is the manufacture of **domestic appliance** equipment. Here a particularly demanding set of requirements applies, as the precoated feedstock needs to be extremely formable, scratch resistant, humidity and temperature resistant, and resistant to domestic cleaners and detergents, as well as aggressive food ingredients. All kinds of large appliance are produced from precoated stock, like refrigerators and freezers, washing machines, dishwashers or ovens, but also smaller items as microwave ovens, ranges or fume hoods (Figure 1.18).

Also, smaller **electrical articles, HVAC** equipment (heating, ventilation, air-conditioning) and **teletronics** are made using prefinished sheet. These items sometimes appear in the

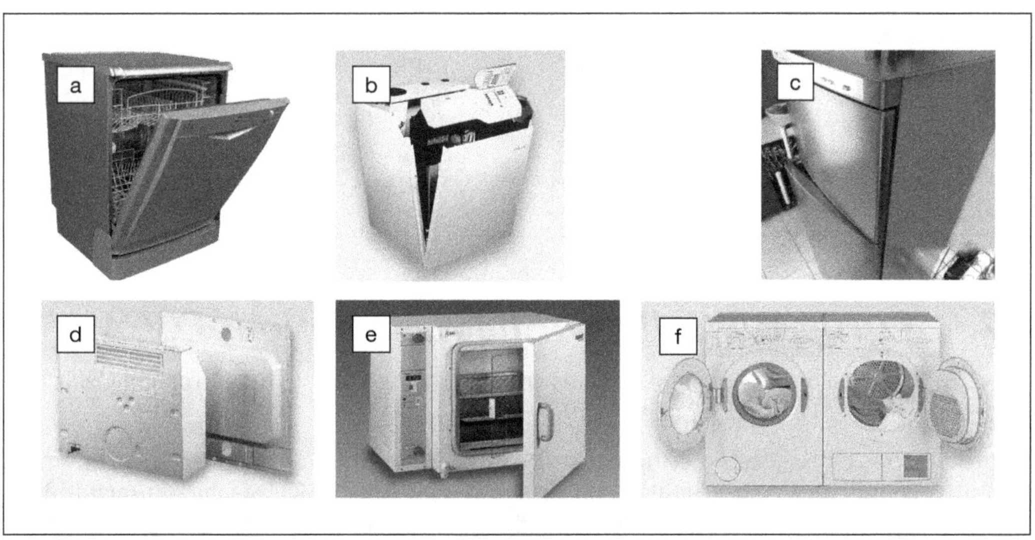

Figure 1.18a to f: Uses of precoated metal in the domestic appliance market Source: a) Tata Steel; b-f) ThyssenKrupp Steel Europe

appliance category, but most of the time they are accounted along with other general industrial uses like metal furniture, lockers, shelves and lighting (Figure 1.19).

An abundance of other possible uses are difficult to categorise, including so different items like metal boxes, roadsign plates, or white display boards.

1.6.5 Packaging applications

Packaging that is related to human or pet food, as mentioned previously, is not covered by the European definition of coil coating (unlike in USA). However a lot of the semi-finished stock that is used for the manufacture of food and beverage cans, can lids and tabs, spray bottles, tubes etc., receives pretreatment and a base lacquer in processes that are identical to coil coating. Figure 1.20 shows some examples of packaging semis that are coated in continuous coil fashion.

Table 1.4 summarises the large variety of possible uses for precoated metal.

Table 1.4: An abundance of uses of precoated metal. Note that in Europe the manufacture of packaging semis for food and beverage cans is not considered as a coil coating segment

Applications of precoated metal

- **Architectural (indoor, outdoor)**
  - Cladding and façades (corrugated, V-shape)
  - Composite insulated panels
  - Doors, garage doors
  - Internal partitioning, ceiling panels
  - Gutters, drains, sinks, flashings
  - Roofing (corrugated, standing seam, 3D-profile)
  - Roller shutters, blinds
  - Structural decking, flooring
- **Transport**
  - Buses, coaches, tankers, trucks
  - Car parts
    - body panels (corrosion protection primer)
    - doors, beams
    - oil filters & sumps
    - instrument dials, wiper blade assemblies, trim
    - license plates
  - Caravans and mobile homes
  - Containers
  - Railway vehicles
- **Appliance**
  - Refrigerators and freezers
  - Washing machines and clothes driers
  - Dishwashers, microwave ovens, cookers
- **General Industry**
  - Lighting assemblies
  - Shelving, furniture, lockers, equipment housing
  - Heating, ventilation, air conditioning
  - Teletronics (Hi-Fi equipment, computers, satellite decoders)
  - Non-stick bakeware
- **Others**
  - Metal enclosures
  - Insulation, cold stores
  - Food industry
  - Accessories
  - Traffic signs
  - Display boards ...
- **Packaging**
  - Beverage and food can-end stock
  - Tabstock

1.7 European Coil Coating Association, ECCA

As an umbrella association for the coil coating industry in Europe, the European Coil Coating Association, ECCA, was founded in 1967.

Its primary task is to promote the use of precoated metal as the outstanding, environmentally sound and cost-efficient construction material. ECCA is headquartered in Brussels, and has 140 members worldwide. In Europe, these include 36 coil coaters, the chemical and paint companies active in this market, as well as the respective manufacturers of machinery, line installations, ovens or measuring devices. In addition, service centres and other metal distributors or stockists that deal with precoated metal are to be found in the members' dictionary. Partner associations and non-European coaters and suppliers are also welcome. Among the membership, there are the large metal and paint industry groups, including steel and aluminium strip producers that operate their own surface finishing plants and sometimes also processing lines for tailored sheet or sandwich panel production. But also smaller groups and medium-sized enterprises are members.

ECCA has organised Marketing and Technical Committees. Each of these performs, through dedicated project teams staffed with delegates from the members, defined tasks for the association. Among these, there are market studies, case studies, collections of educational material, definitions on quality and design of precoated metal, environmental studies, etc.

There are two major annual events, the Spring Conference and the Autumn Congress. During these events, member delegates or external experts present technical and marketing papers, the Committees and Management do their reports, and association affairs are dealt with and communicated.

Table 1.5: Key data and selected major activities of the European Coil Coating Association, ECCA

European Coil Coating Association (ECCA)

- Brussels (founded 1967), www.prepaintedmetal.eu
- 140 members worldwide
  - 36 coil coaters in Europe
    - Steel and aluminium plants with integrated surface finishing, partially onward processing (sheet lines, sandwich panels ...), medium-sized enterprises
- Suppliers: Chemical and paint industry
- Machinery and equipment manufacturers
- Service Centres
- Partner Associations

Selected ECCA activities

- Marketing and Technical Committees, with project teams
- Annual Spring Conference, Autumn Congress
- Shipment statistics
  - Products (steel, aluminium)
  - Paint demand
  - Target markets
  - Delivery flows
- Support/cooperation in European standardisation (CEN) and environmental legislation, sustainability reporting
- Outdoor exposure site programme Eurodes
- Cooperation with NCCA (National Coil Coating Association), Cleveland OH, USA; www.coilcoating.org

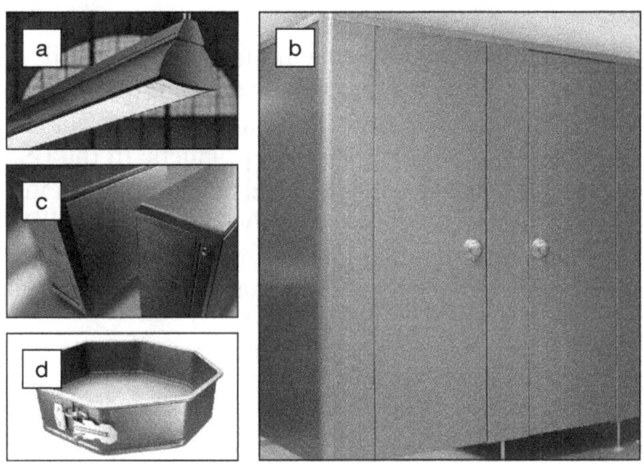

Figure 1.19a-d: Uses of precoated metal for general industrial purposes Source: a-c) ThyssenKrupp Steel Europe; d) Tata Steel

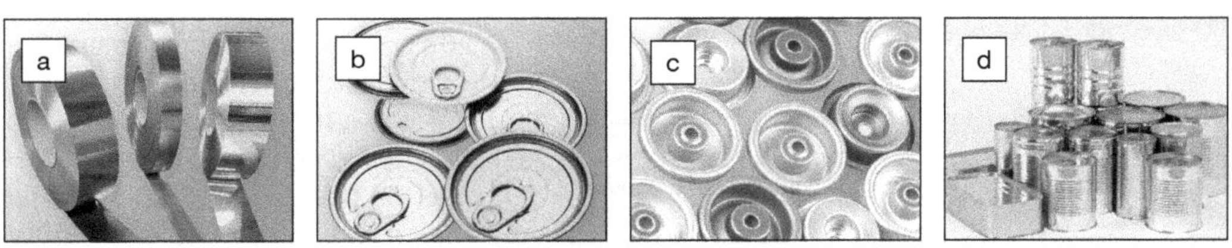

Figure 1.20a to d: Packaging semis made from precoated metal Source: ThyssenKrupp Rasselstein

ECCA collects and compiles statistical data about the aspects of the members' business. These data are anonymised for confidentiality reasons, and published quarterly. They deal with the sales development of the products, their distribution to the target markets by region and application, developments in the flow of goods delivered, and the paint demand.

Another important activity is the support work for the European standardisation, and environmental legislative bodies.

ECCA runs its own test sites for outdoor exposure, the Eurodes programme, and cooperates with specialist institutes for corrosion research.

A close liaison is maintained with the American National Coil Coating Association, the only other organisation of comparable scope in this worldwide market.

Finally, education about the subject is of paramount importance, and a good tool to make the case for precoated metal. Learning of what coil coating is all about and what quality and design opportunities can be achieved with it, is equally important for interested users or suppliers, for architects, engineers and designers, for students of these subjects, but also their professors. Therefore, on its website, www.prepaintedmetal.eu, ECCA offers an interactive learning programme, the ECCA Academy. Everyone can register to it free of charge. The programme has been put together by experts from ECCA members and covers all the subjects that are of relevance for coil coating. At the end of each module of the training programme, an interactive exam can be taken. Having passed all exams, a Certificate can be obtained.

1.8 History of coil coating

The origin of coil coating [7-9] can be dated back to the turn of the 19th century in Germany, where the concept of paint coating of narrow steel strips was first realised. Shortly before WW I, coated strip steel ("Lackiertes Bandeisen") up to 350 mm width appeared in contemporary advertisements of medium-sized steel mills like *Trierer Walzwerke* and *Rheinische Stahlwerke*. While the technology proliferated abroad in the 1930s and 1940s, narrow steel and aluminium strip lines were put on stream, mainly for the production of stock for Venetian blinds and for niche market products. In the USA broad strip lines (> 600 mm width) emerged in the 1960s in both steel and aluminium industries. The market quickly extended throughout Europe (Belgium, Italy, Sweden, and UK), North America as well as Japan. Also lines for aluminium foil lamination were first erected in those days. Further development saw lines with increased dimensions and speeds, as well as combined lines with coating sections following galvanising or annealing stages.

Table 1.6: Coil coating - a lifetime of development Source: Mitchell[9]

	1940s	Present
Line speed (m/min)	12	120
Substrate	Aluminium	Al, Z, ZE, ZA, AZ, CRS
Paint systems	Alkyds, solution vinyls, acrylics	Polyesters, PVdF, PVD, polyurethane, epoxies, powder
Market applications	Furniture, building	Building, appliance, HiFi, HVAC, auto, furniture, bakeware, packaging

Parallel to the general technology, the pretreatment processes developed, elaborating cleaning as well as conversion coatings like acid chromate solutions, formulated chromates, phosphochromates, to phosphate, alkaline, and finally no-rinse (dry-in-place) coatings. Recently, chromium-free processes have widely replaced the hazardous chromate containing processes (cf. Chapter 4). Also the variety of substrates and paint chemistries has spread grossly, making coil coated metal available for an equally wide variety of end uses (Table 1.6). Nowadays, coil coating has grown to a large, worldwide industry. The saturated markets in Europe, North America and Japan are being complemented by the quickly emerging coil industry in China and India.

1.9 Literature

[1] Meuthen, B., Jandel, A.-S., Coil Coating, 2nd ed., Vieweg, Wiesbaden 2008, pp. 1ff

[2] Sander, J., The Basics of Coil and Can Coating, Tutorial, Europ. Coatings Conf. Coil and Can Coatings 2011, Vincentz Network, Hannover 2011

[3] anon., ECCA Academy, European Coil Coating Association (ECCA), Brussels 2011, www.prepaintedmetal.eu

[4] Meuthen, B., Bandbeschichtung (Coil Coating), Seminar "Compact Coil Coating", Proc., Techn. Akad. Wuppertal 2011

[5] Hauchard, E., ECCA Statistics 2012, ECCA General Meeting 2013, Proc., European Coil Coating Association, Brussels 2013

[6] Bielefeld, F.W., Sander, J., The Profitable Choice: From small-batch post-painting to Compact Coil Coating, European Coatings Market Day Coil and Can Coatings 2011, Proc., Vincentz Network, Hannover 2011

[7] Meuthen, B., 100 Jahre bunter Stahl: Die Geschichte der Bandlackierung in Deutschland (The History of Coil Coating in Germany), Forum Session on the Biennial of the Hoesch Museum, ThyssenKrupp, Dortmund 2007

[8] Meuthen, B., Jandel, A.-S. [1], pp. 5f

[9] Mitchell, P.J., The History And Development Of Coil Coating Pretreatments, ECCA Autumn Congress 2009, Proc., European Coil Coating Association, Brussels 2009

2 Corrosion protection

2.1 Principles of function

2.1.1 Corrosion phenomena [1]

Worldwide, approx. 5 % of the annual tonnage of produced metal is lost to corrosion. This is due to how easily the most common construction metals are oxidised, thereby forming their oxides, and annihilating the high effort of production. For example, the generation of metallic iron from its ore (cf. also Chapter 4.2.2), as shown in Equation 2.1, consumes 138.2 kJ/mol of energy. This energy is stored in the reaction products, and can be regained by the corrosion reaction that therefore occurs spontaneously.

Equation 2.1:

$$\begin{array}{lcl} 172.6\ \text{kJ} + CO_2 + C & \rightarrow & 2\ CO \\ \underline{2\ FeO + 2\ CO} & \underline{\rightarrow} & \underline{2\ Fe + 2\ CO_2 + 34.4\ \text{kJ}} \\ \mathbf{138.2\ kJ} + 2\ FeO + C & \rightarrow & 2\ Fe + CO_2 \end{array}$$

It is estimated that a quarter of such losses can be avoided by appropriate protection measures. Corrosion occurs where buildings, technical installations and equipment are subject to weathering, or in contact with soil, sweet or salt/brackish water.

Several forms of corrosion are known. Some of these are summarised in Table 2.1 [2].

Obviously, the degree of corrosion and the speed of its propagation depend on the aggressiveness of the surroundings a building or piece of equipment is exposed to. The conditions of environments have been classified according to the degree of corrosion they typically cause. Table 2.2 gives the corrosivity classification of environments according to the relevant standard, EN ISO 12944-2 [3].

Figure 2.1: Delamination of paint from a galvanised steel surface, caused by corrosion

Table 2.1: Corrosion phenomena

Corrosion	Corrosion phenomena
Galvanic corrosion	Occurs upon galvanic contact between metals of different electro-negativity, e.g. brass screw and Al sheet; Al rivet and carbon steel joined. Both metals act as electrodes, exchanging a weak electric current; the less noble metal serves as anode, and corrodes (dissolves); oxygen reduction takes place at the nobler metal (cathode). An electrically conductive medium (moisture) is a prerequisite.
Crevice corrosion	Occurs in crevices between metallic surfaces, e.g. in thread gaps, where concentration gradients in the electrolyte inside and outside the crevice cannot be levelled out by diffusion, due to the narrowness of the gap.
Filiform corrosion	Thread-forming corrosion underneath organic coatings, starting from a failure spot, scratch, etc. Typically, it occurs at a relative humidity below dew point (40 to 85 % rH). The corrosion threads grow along grain boundaries, and never cross each other.
Pitting corrosion	By local damage of the passivation layer and the resulting formation of local elements, e.g. Al/stainless steel. May lead to complete perforation of structures (pipes, containers).
Stress-fatigue corrosion	Propagating formation of fissures in metals by simultaneous action of a corrosive medium and a strain resulting in fast degradation of the material, may cause sudden structural damages.

Table 2.2: Corrosivity classification of environmental conditions according to EN ISO 12944-2 [3]

Corrosivity class	Gauge loss first year [µm]		Examples of typical environments	
	Carbon steel	**Galvanised**	**Outdoor**	**Indoor**
C1 negligible	<1.3	<0.1	–	≤60 % rel. humidity
C2 low	1.3 to 25	0.1 to 0.7	Low-polluted atmosphere, dry climate	Uninsulated bulding with temporary condensation
C3 medium	25 to 50	0.7 to 2.1	Urban/industrial atmosph. with moderate SO_2 load, or moderate seashore climate	Rooms with high relative humidity and low pollution
C4 strong	50 to 80	2.1 to 4.2	Industrial atmosph., or coast with moderate salt load	e.g. chemical production, swimming baths
C5 very strong I	80 to 200	4.2 to 8.4	Industrial atmosph., with high rel. humidity and aggressive atmosphere	Buildings with virtually constant condensation and high pollution
C5 very strong M	80 to 200	4.2 to 8.4	Coastal and offshore areas	

2.1.2 *Strategies for corrosion inhibition*

As corrosion is driven by the formation of electrochemical couples, any prevention must see to avoid the direct contact of materials, especially metals, with different electronegativity. Furthermore, it is required to suppress or sufficiently retard the reactions that take place at the metal surfaces, rendering the electrode sites for the corrosive event.

For atmospheric corrosion, these reactions can be represented by Equations 2.2 and 2.3.

Equation 2.2: $Zn \rightarrow Zn^{2+} + 2\,e^-$ Anodic reaction

Equation 2.3: $1/2\,O_2 + H_2O + 2\,e^- \rightarrow 2\,(OH)^-$ Cathodic reaction

Upon its dissolution, the metal releases free electrons, which in turn are consumed for the reduction of oxygen. Figure 2.2 shows a schematic view of this situation [4-6].

Common strategies employed [6, 7] for corrosion prevention are:

- Conversion coatings
- Cathodic protection
  - Galvanising
  - Zinc dust coatings
- Sealing of pores
  - Multilayer paint coatings
- Barrier effect
- Dielectric features of a coating
- Reduced absorption of humidity/gas
- Interception of corrosive agents
  - Active anions
  - Ion exchanger pigments
- Improvement of the substrate adhesion
  - Adhesive primers
  - Silane/siloxane coatings
- Increase of alkalinity
  - Concrete coatings

Humidity plays a major role in adhesion failures and corrosion. The adherence of organic coatings generally decreases when penetrated by aqueous media, through swelling,

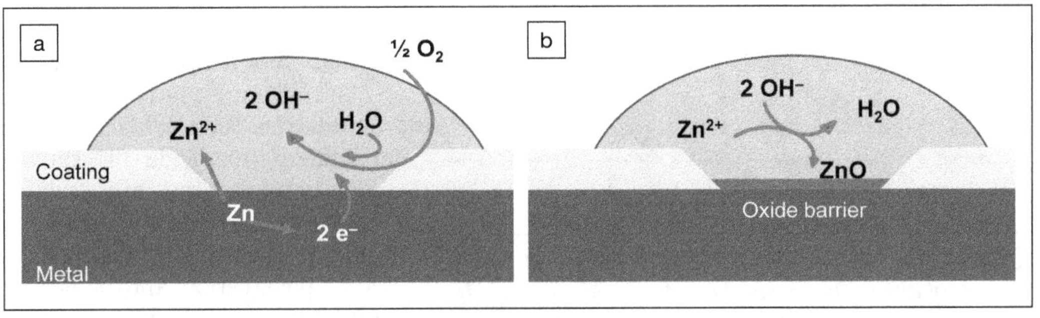

Figure 2.2: a) Atmospheric corrosion at a punctured coating over a zinc substrate; b) Formation of a barrier layer by precipitation of zinc oxide (native oxide barrier)

softening, hydrolysis, and cleavage of hydrogen bonds. Moreover, the presence of water in the interface enables the diffusion of electrolyte, and hence ion transport. Thus, delamination of an organic coating is also a result of corrosion. Figure 2.3 illustrates the so called cathodic disbonding process that propagates along the interface between the metal and the coating. As the reduction of atmospheric oxygen renders an alkaline environment by the formation of hydroxyl ions (OH^-), oxidic conversion layers that are soluble in alkali may be damaged.

2.1.3 *Metal oxide formation, passivation and conversion coating*

Native oxide coatings that form a barrier layer decelerating the corrosion process, develop on e.g. zinc (Zn), aluminium (Al), magnesium (Mg), or chromium (Cr). Also carbonate ions from atmospheric carbon dioxide may be incorporated in these layers (cf. Chapter 3.3.3.1). The oxide films are particularly effective when they are insoluble in an alkaline environment as is formed during the oxygen reduction. For example, over Zn-Mg coatings, the corrosion and the related current density drops to zero for lower (more negative) electrode potentials and higher alkalinities, the more Mg is present in the metallic coating (cf. Figure 2.4) [8].

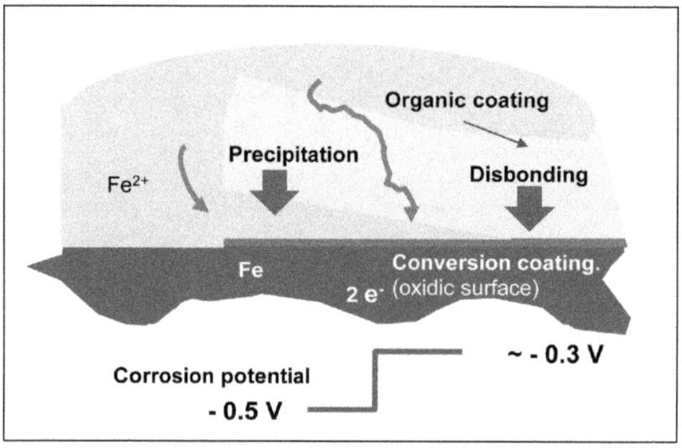

Figure 2.3: Schematic representation of the cathodic disbonding mechanism

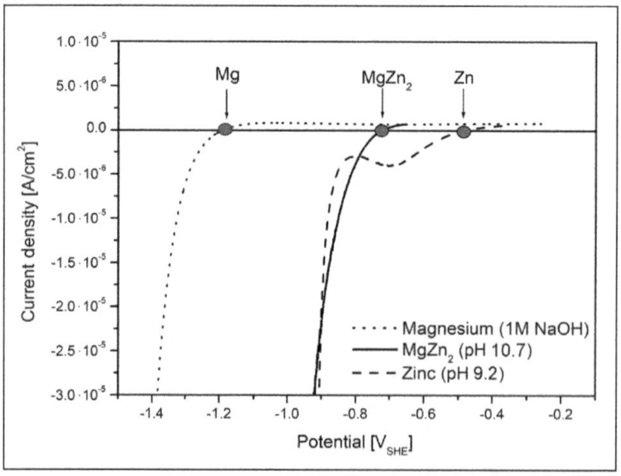

Figure 2.4: Current/potential curves of passive electrodes in aerated solutions. The passive layers of Zn, $MgZn_2$ and Mg consist of ZnO, a mixed Zn-Mg hydroxide ([Mg/Zn] = 0.2) and $Mg(OH)_2$, respectively Source: Elsevier Ltd. 2009; cf. Hausbrand et al. [8]

Even though such oxide barriers display a good stability against atmospheric corrosion, the presence of chloride ions is critical. By the displacement of oxygen from the crystal lattice, oxy/hydroxy chloride species are generated that induce local (pitting) corrosion. Hence, the incorporation of substances that can intercept chloride ions is a common approach to enhance the corrosion resistance of a coating [9, 10].

In industrial painting, instead of relying on a native oxide layer, the formation of conversion layers in a controlled pretreatment process is preferred. Such layers are water insoluble in a wide pH range. They provide adhesion between the metal substrate and the paint, and they serve as physical barriers, passivation and/or active inhibitors against intruding corrosive substances.

2.1.4 Cathodic protection

A common method of corrosion protection of ferrous materials involves their coating with a layer that contains a less noble metal, e.g. zinc (Zn). The latter has a lower electrochemical potential (E_{pot} = - 0.76 V) than the ferrous substrate (E_{pot} = - 0.44 V). Therefore, it corrodes preferentially, keeping the base metal from destruction. This effect is usually described by the terms "sacrificial anode", and "cathodic protection". Zinc pigmented primers, quite common anti-corrosion barriers in artisan and industrial batch coating [11-14] are also applied by coil coating processes on sheet for the car manufacture (corrosion protection primers, CPP) [15, 16]. They provide enhanced corrosion protection in critical areas of the car body, like seams, box sections or flanges, where pretreatment and cathodic electro-dip primer of the body shop operation are difficult to apply.

2.2 Design of organic coating systems

2.2.1 Diffusion barriers

Despite proper film formation and cross-linking, no polymer film is entirely impermeable to gases or electrolytes. The random development of pores and crevices in the film, as is caused by the relative thermal mobility of polymer moieties [17], inadvertently enables diffusion and migration to some extent. This leads to water uptake, swelling and plasticising of the coating, [18] and penetration of dissolved ions that may cause chemical degradation of the film and corrosion of the substrate [19].

Film thickness increases, multiple layer coatings, and the choice of film builders with a low water uptake are possible strategies to enhance the barrier effect of an organic coating. Coil coating primers, for example, are usually applied with 5 µm dry-film thickness. However, they may be extended up to 30 µm under particularly severe conditions, like for seashore structures [20].

Resins like epoxies or polyurethanes show only little water uptake in a humid atmosphere, but may break down in condensation environments. Blending, grafting, or copolymerisation render more suitable resin films with a moderate hydrophobicity, hence also less water absorption [21]. However, the polymer design must accommodate an entire set of sometimes conflicting requirements, regarding e.g. formability, adhesion, curing temperature, diffusion barrier properties, chemical resistance or wetting.

2.2.2 Pigments

Active pigments are introduced that react with penetrating corrosive species (cf. also Chapter 5.7). Strontium chromate and dichromates, widely used in the past, are currently replaced by less toxic successors.

The mechanism effective in commercial calcium-modified silicates involves the release of calcium ions (Ca^{2+}) under acid conditions via an ion-exchange [22].

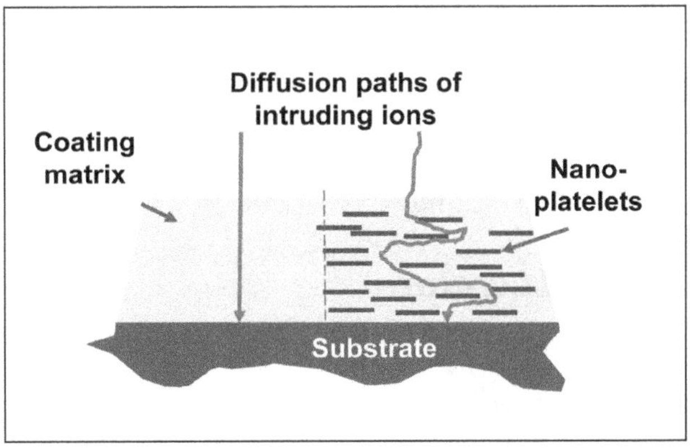

Figure 2.5: Schematic view of the extension of the diffusion path by a nano-platelet filled coating Illustration according to Lewis [24]

Table 2.3: Pigment types used in corrosion prevention

Pigment types	Anti-corrosion mechanism
Active pigments	Inhibition of active corrosion events Electrochemically Chemically
Barrier pigments	Retardation Obstruction of the transport of corrosive substances
Sacrificial pigments	Cathodic protection Sealing of pores by cementation

The calcium ions reduce the corrosivity of the electrolyte, and precipitate corrosive chloride and sulphate ions.

Mineral or synthetic hydrotalcites or zeolites have been employed as carriers for anti-corrosive ions like molybdates or vanadates. The release of these active ions can be controlled by the conditions of the surrounding medium [23].

Natural or synthetic minerals packed of agglomerated crystalline platelets are used to extend the diffusion path for corrosive ions or media through the coating, cf. Figure 2.5. This effect has been observed for oxygen, carbon dioxide and water vapour [24].

Cathodic protection is provided by sacrificial pigments like zinc (Zn) that, upon dissolution, form an additional barrier through onward reaction of the dissolved species and precipitation of the reaction products.

2.3 Function of individual coating layers [25, 26]

An industrial coating system on metal usually is made up of several layers. A basic conversion coating first covers the metal with an anticorrosive barrier and provides the primary adhesion between the substrate and the organic coating.

In coil coating, most commonly, two-coat paint systems are used, i.e. a primer and a topcoat layer are applied successively. The primer gives additional corrosion protection and mediates between the different mechanical and thermal features of the substrate and the overlying topcoat. This is achieved by strong bonding to both neighbouring materials and by providing a flexible layer that allows the distribution of mechanical stress by deformation and thermal expansion and contraction. Chemical resistance requires the crosslinking of single polymer strands, and corrosion resistance is brought on by suitable pigmentation.

The topcoat renders the overall chemical and mechanical features by appropriate choice of the resin and crosslinking (cf. Chapter 5), as well as the aesthetic appearance (colour and gloss). Figure 1.8 (see Chapter 1.2) illustrates the layers of a typical coil coating paint system, their individual functions and dimensions.

For applications in particularly demanding corrosive environment, also coatings with a higher total dry-film thickness are common, be it three-coat layers with intermediate coats (e.g. 30 to 80 µm), PVC plastisol coatings (100 to 200 µm), or colaminates with permanent foils. On the other hand, also single paint coats are used where appropriate, like for coatings on Al, interior coatings and backers.

2.4 Literature

[1] Sander, J., Corrosion Protection Coatings, European Coatings Seminars, Amsterdam 2012, Proc., Vincentz, Hannover 2012

[2] anon., Corrosion Fundamentals, NASA, Kennedy Space Center 2013, http://corrosion.ksc.nasa.gov/galcorr.htm (18-06-13, 18:35 h)

[3] DIN EN ISO 12944-2:1998-07, Beschichtungsstoffe - Korrosionsschutz von Stahlbauten durch Beschichtungssysteme - Teil 2: Einteilung der Umgebungsbedingungen (Paints and varnishes -

Corrosion protection of steel structures by protective paint systems - Part 2: Classification of environments) (ISO 12944-2:1998), Beuth, Berlin 1998

[4] Meuthen, B., Jandel, A.-S., Coil Coating, 2nd ed., Vieweg, Wiesbaden 2008, p. 89 ff

[5] Fafilek, G., Kronberger, H., Thermodynamik, metallische Werkstoffe und Korrosion (Thermodynamics, Metallic Construction Materials and Corrosion), Experimental Exercises Script, Exercise TK, Inst. Chem. Technologies and Analytics, TU Vienna 2005, p. 7; b) Evans, U.R., The action of salt solution on iron and steel, in the presence of oxygen, J.Chem.Soc. 44, 1924, pp. 315 ff, cit. re. Fafilek, G., priv. comm., TU Vienna 2013

[6] Lewis, O. D., A Study of the Influence of Nanofiller Additives on the Performance of Waterborne Primer Coatings, PhD Thesis, Loughborough Univ., Loughborough 2008, p. 78

[7] Vogelsang, J. A., Selbstheilende Beschichtungssysteme - Ein Überblick (Self-Healing Coating Systems...), Schiff und Hafen (8) 2009, pp. 28 ff

[8] Hausbrand, R., Stratmann, M., Rohwerder, M., Corrosion of Zinc-Magnesium Coatings: Mechanism of Paint Delamination, Corr. Sci. 51, 2009, pp. 2107 ff

[9] Mahajanam, S. V. P., Buchheit, R. G., Characterization of Inhibitor Release from Zn-Al-$[V_{10}O_{28}]^{6-}$-Hydrotalcite Pigments and Corrosion Protection from Hydrotalcite-Pigmented Epoxy Coatings, Corr. 64, 2008, pp. 230 ff

[10] Crass, G., Janocha, S., Bothe, L. (inv.), Biaxially oriented multilayered polyolefin film which is printable and of which both sides are sealable, its production and use, European Patent EP 0263963, Hoechst AG 1987

[11] Meyer, K., Schutz von Stahlkonstruktionen Beispiele aus der Eisenindustrie und dem Bergbau (Protection of Steel Constructions ...), Fette, Seifen, Anstrichmittel 69, 1967, pp. 90 ff

[12] Schröder, T., Erfahrungen mit Schutzbeschichtungen an Stahlwasserbauten der Wasser- und Schiffahrtsverwaltung im Bereich der Nord- und Ostsee (Experiences with Protective Coatings on Hydraulic Structures...), Materials and Corrosion 23, 1972, pp. 993 ff

[13] anon., Kanalüberführung Leine (Canal Flyover Leine River), Wasser- und Schifffahrtsverwaltung des Bundes - Neubauamt Hannover (Federal Waterways and Shipping Administration), Hannover 2009; www.nba-hannover.wsv.de/baumassnahmen/abgeschlossene_baumassnahmen/neubau_leinequerung.html (26-06-13; 18:36 h)

[14] Vogelsang [7], p. 30

[15] Meuthen, Jandel [4], p. 75 ff

[16] Schinzel, M., Advanced Corrosion Protection of Automotive Body Sheet - A Challenge for Coil Coating, ECCA Autumn Congress, Proc., European Coil Coating Association, Brussels 2009

[17] Lewis [6], p. 87

[18] Goldschmidt, A., Streitberger, H.-J., Handbuch Lackiertechnik (Handbook Coating Technology), BASF Coatings AG, Münster 2002, p. 447

[19] Goldschmidt, Streitberger [18], p. 401

[20] Sander, J., Kirmaier, L., Manea, M., Shchukin, D., Skorb, E., Anticorrosive Coatings, Vincentz Network, Hannover 2010, p. 10

[21] Öchsner, W. P., Bergk, B., Fischer, E., Gaszner, K., Sorptionsisothermen für Wasser in organischen Beschichtungen und deren Einfluss auf die Beschichtungseigenschaften (Sorption Isotherms for Water in Organic Coatings ...), Farbe und Lack 111, 2005, pp. 42 ff

[22] Vogelsang [7], p. 29

[23] Mahajanam, Buchheit [9], p. 231

[24] Lewis [6], p. 34

[25] Kittel, H., Streitberger, H.-J. (ed.), Lehrbuch der Lacke und Beschichtungen (Coursebook of Paints and Coatings), Vol. 6, 2nd ed., Hirzel, Stuttgart 2008, pp. 485 ff

[26] Meuthen, Jandel [4], pp. 51, 79 ff

3 Industrial cleaning

3.1 Introduction

An organic coating on a metallic substrate requires a properly clean and pretreated metal surface. High durability and a long-lasting, appealing aspect of any coating substantially depend on the quality of the surface preparation.

While mechanical cleaning can provide a basic level of surface cleaning, in coil coating, chemical cleaning is a necessity. The coil surfaces are most efficiently and economically cleaned by aqueous spray cleaners, but also immersion cleaning is common. The process must be adjusted to the substrate and the type of contamination in either case.

The difference between a clean and a contaminated surface often is not apparent to visual inspection. To ensure that both a perfect pretreatment and organic coating is achieved, the cleaning stage therefore must be carefully monitored and maintained [1, 2].

Typical contaminations on feedstock for coil coating are:

- Rust, scale, oxide layers
- Oil, fats, waxes, rolling oils, emulsifiers, corrosion protective oils
- Fines, abrasions, dirt

Oxidic and oily films reduce the surface tension, hence adversely affect any subsequent water-based process.

3.2 Why cleaning?

3.2.1 Contaminations

Industrial cleaning deals with a great variety of soils and contaminants. Even though feedstock for a coil coating process is usually high-quality material that is generated in a controlled process and is subject to entry inspection, still contaminations are commonly encountered (see Chapter 3.1).

Rust and scale may require mechanical precleaning or pickling.

Oils and fats usually originate from the generation process of the material, e.g. rolling oils, protective oils (often containing additional organic inhibitors, e.g. triethanol amine), not to forget contaminations from manual handling, i.e. sweat or fingerprints that contain fats, fatty acids, proteins etc.

Fines and metal abrasions originate from rolling, cutting or slitting processes.

Though the metal surface might look clean, oxidic and oily films obscure any onward process in the course of a coating operation, in particular when aqueous media are involved.

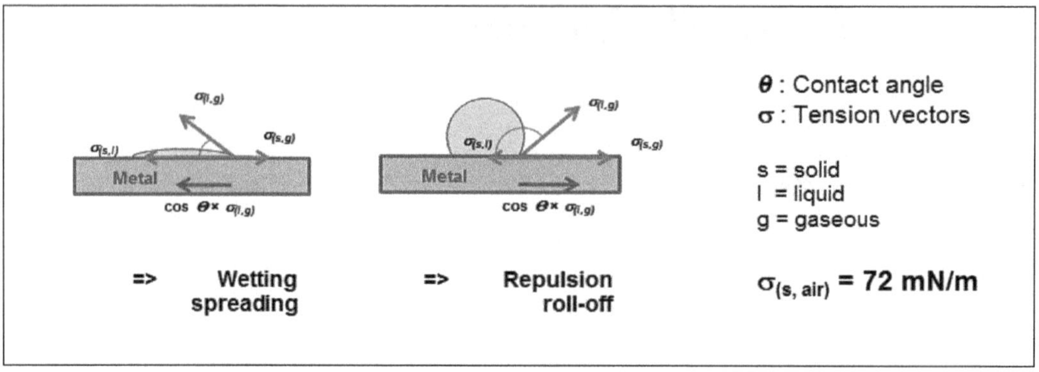

Figure 3.1: Schematic representation of tensions active at the interface of a droplet on a metal surface

3.2.2 Cleanliness and surface tension [3]

The complete wettability of the metal is often used as an indicator of "chemical cleanliness". One should be aware, however, that water-break-free rinsing may be affected by the presence of surfactants, e.g. in self-emulsifying oils, the thickness of the water film, and the evaporation.

A measure for the wettability of a surface is its surface tension [3]. It results from differences of the attractive forces that apply to the molecules inside versus between non-miscible phases. The form of droplets, micelles and foam bubbles, the adsorption of gases to solids, capillarity phenomena like the wetting and spreading of liquids on a metal surface are all manifestations of surface tension [4].

A mathematical description is given by Young's equation. The wettability of a solid by a liquid is characterised by the contact angle Θ that a drop of the liquid adopts on the solid surface (Figure 3.1).

The most common method for the measurement of surface tension uses the direct, microscopic measurement of a sessile drop. Figure 3.2 shows an optical instrument for this method. A droplet is automatically placed on the adjustable table on the right of the picture, and inspected by the CCD camera on the left. Wetting is indicated by contact angles $< 90°$ (i.e. $\cos \Theta > 0$), repulsion by angles $> 90°$ ($\cos \Theta < 0$). Good wetting is obtained when the interfacial tension between the solid and the liquid is higher than the tension at the solid-gas boundary. The interfacial tension should exceed 72 mN/m, the surface tension of water versus air, if the following process involves wetting of the surface by aqueous media [5].

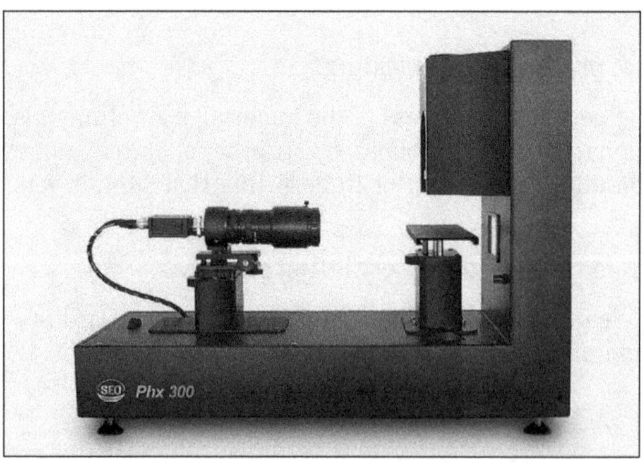

Figure 3.2: Camera system for the microscopic determination of contact angles at a sessile drop Source: Elicomarketing

For the determination of the surface tension of a solid surface, the contact angles of droplets of different liquids with known surface tensions are measured and the resulting overall surface tension of the solid calculated. Other methods involve measurements of the forces that apply when a liquid is deflected by capillarity, i.e. the meniscus of a liquid in a capillary tube, or weighing of droplets that drip off a capillary tip. A simple method for measuring surface tension uses prefabricated test inks [6].

Table 3.1: Adhesion promoting forces prevailing in paint and film coatings

Type of attraction force	Range [nm]	Bonding energy [kJ/mol]
van-der-Waals force, dipolar/inductive forces	1	5 to 25
Hydrogen bonds between metal/ oxide surfaces and protic molecules (–OH, –SH, –NH–)	0.1 to 1	~50 kJ/mol
Chemical bonds, e.g. metal–oxygen	0.1	>>100 kJ/mol

3.2.3 *Paint adhesion* [7]

Any organic coating must adhere to its underground despite any deformation, shear or torque force during metalworking, or dynamic load or also accidental deformation during service life. Flanging, profiling, or bending operations, cutting, drilling, honing, or screw fastening, vibrations and torsion or impact and indents are examples for the mechanical stress a work piece may suffer from. Mechanical stress also arises from hydrothermally induced volume changes of coatings and substrates.

Adhesion of an organic coating is a precondition for a long lasting coated product to serve in a corrosive environment. In particular for coil coating, where extreme deformations of the sheet metal are applied to shape the final work piece (e.g. domestic appliance equipment housing, car body parts made from preprimed sheet, or food cans and lids and crown corks), insufficient adhesion and elasticity may cause cracks, if not immediate delamination, that form inroads for corrosion attack.

One keeps hearing the negligent simplification that paint sticks well on a metal surface as long as this is sufficiently rough and porous. The paint might then seep into the troughs and clamp especially strongly. However, though a certain roughness of the surface is helpful, coil coating, due to the low dry-film thickness, does not allow substrate roughness of a micrometer or more, as is obtained by grinding or blasting. Moreover, adhesion is much more due to van-der-Waals forces, and dipole and inductive attraction between the substrate and molecules with a binding energy level of 5 to 25 kJ/mol. More strongly, hydrogen bonds form between protic molecules (hydroxy, mercapto, amino functions) and metallic or oxidic surfaces, representing the 50 kJ/mol level. Chemical bonds like metal-to-oxygen bonds are even stronger, i.e. several 100 kJ/mol, hence desirable for adhesion promotion and reliable, long-lasting protection against creepage (Table 3.1). Therefore, undisturbed wettability that allows interaction in the nanometer scale is an imperative, and it can only be achieved by proper cleaning and removal of all substances and separating layers.

3.3 *Aqueous cleaning* [8]

3.3.1 *Overview*

Aqueous immersion or spray processes are used in most industrial metal coating lines, to clean the metal surface, and remove oil, solid contaminations and superficial scale. Only on steel coil, additional mechanical cleaning is used to support the removal of solid soil and rust by intermediate brushing between aqueous pre- and main cleaning stages.

Typically, alkaline aqueous cleaners are used on steel and galvanised steel substrates, while for aluminium, also acidic cleaning is common. Both are sufficient to remove also the native oxide films.

Table 3.2: Ingredients of industrial cleaners

Type of cleaners	Ingredients
Acidic cleaners	Phosphoric acid Dihydrogen phosphates (Al also: Sulphuric acid) Fluorides, fluoro complexes Surfactants, solubilisers
Alkaline cleaners	Alkali hydroxide Phosphates, carbonates, silicates Surfactants, solubilisers Complexants, sequestrants Inhibitors
Neutral cleaners	Surfactants, solubilisers Inhibitors

When cleaning aluminium with an alkaline cleaner, the surface is left with a layer of insoluble alkaline, oxidic compounds that would quench subsequent acidic treatment baths. Therefore, it is common practice to remove these alkaline residues with an intermediate, acidic rinse or deoxidiser. As aluminium is amphoteric, it is also dissolved in acids. Therefore, acid cleaning is common practice whenever a low contamination level allows (coil, can).

Finally, also neutral cleaners might be used for substrates that are especially critical to etching. Table 3.2 summarises the typical ingredients of each of these cleaner classes.

3.3.2 Alkaline cleaners

Alkalis are employed in most aqueous cleaners to saponify fats and oils, and to pickle the metal surface. Builders (e.g. phosphate, silicate) disperse solid dirt particles in the solution after their removal from the surface. Surfactants ensure quick wetting of the metal surface and emulsification of fatty and oily contaminations in the cleaner bath. Moreover, additives like defoamers often are present.

The alkalinity of the cleaner changes during the course of reaction. This is mainly due to the consumption of the alkali by saponification and metal dissolution. Also, ageing effects caused by carbonation through entrapment of carbon dioxide from the air must be accounted for. Hence, the alkalinity must be carefully monitored by manual or automatic methods, and balanced by replenishing the cleaner chemical whenever necessary.

3.3.3 Mechanism of alkaline cleaning, bath age and control

The chemical dissolution reaction of the oxide film, stage A (Equations 3.1 to 3.3), and the electrochemical dissolution of the substrate, stage B (Equation 3.4 to 3.5), are depicted further.

Stage A: Dissolution of the native oxide film

Equation 3.1: $ZnO + 2\ OH^- + H_2O \rightarrow (Zn[OH]_4)^{2-}$

Equation 3.2: $Zn(OH)_2 + 2\ OH^- \rightarrow (Zn[OH]_4)^{2-}$

Equation 3.3: $Zn(OH)_6(CO_3)_2 + 14\ OH^- \rightarrow 5\ (Zn[OH]_4)^{2-} + 2\ (CO_3)^{2-}$

Stage B: Dissolution of zinc

Equation 3.4: $Zn + 4\ OH^- \rightarrow (Zn[OH]_4)^{2-} + 2\ e^-$ Anodic reaction

Equation 3.5: $O_2 + 4\ e^- + H_2O \rightarrow 4\ OH^-$ Cathodic reaction (inhibited by oxygen depletion)

Carbonation

Equation 3.6: $2\ OH^- + CO_2 \rightarrow (CO_3)^{2-} + 2\ H_2O$

Saponification of oils and fats

Equation 3.7:

$$\begin{array}{ll} H_2C\text{-}O\text{-}(CO)\text{-}(CH_2)_n\text{-}CH_3 & \\ \quad | & \\ HC\text{-}O\text{-}(CO)\text{-}(CH_2)_n\text{-}CH_3 + 3\ OH^- & \text{Triglyceride (fat) + alkali} \\ \quad | & \\ H_2C\text{-}O\text{-}(CO)\text{-}(CH_2)_n\text{-}CH_3 & \end{array}$$

$$\begin{array}{ll} \quad H_2C\text{-}OH & \\ \qquad | & \\ \longrightarrow HC\text{-}OH + 3\ H_3C\text{-}(CH_2)_n\text{-}(CO_2)^- & \text{Glycerol + carboxylate (soap)} \\ \qquad | & \\ \quad H_2C\text{-}OH & \end{array}$$

These reactions were studied on galvanised surfaces, using an electro-chemical quartz crystal micro-balance (ECQM) array. Figures 3.3a and b illustrate the experimental setup and the graph displaying the major result of the study [9]. The experimental results confirmed the two different reactions that correspond to the chemical equations given above:

Stage A: Dissolution of the native oxide/carbonate/hydrate layer (chemical), Stage B: Dissolution of metallic zinc and zinc hydroxide layer formation (electrochemical).

When oxygen is absent or substantially depleted, as may happen underneath an organic coating, the cathodic reaction is inhibited, so that only chemical dissolution is observed. Equation 3.6 also shows the consumption of free alkali by the carbonation reaction. Finally, alkalinity is also consumed by saponification of oils and fats. The carboxylate soap thereby rendered (Equation 3.7) acting as a surfactant, hence may cause foam generation. The dissolution of Zn is thus influenced by aeration, but also by the alkalinity, the build-up of dissolved zinc ions and carbon dioxide absorption from the air.

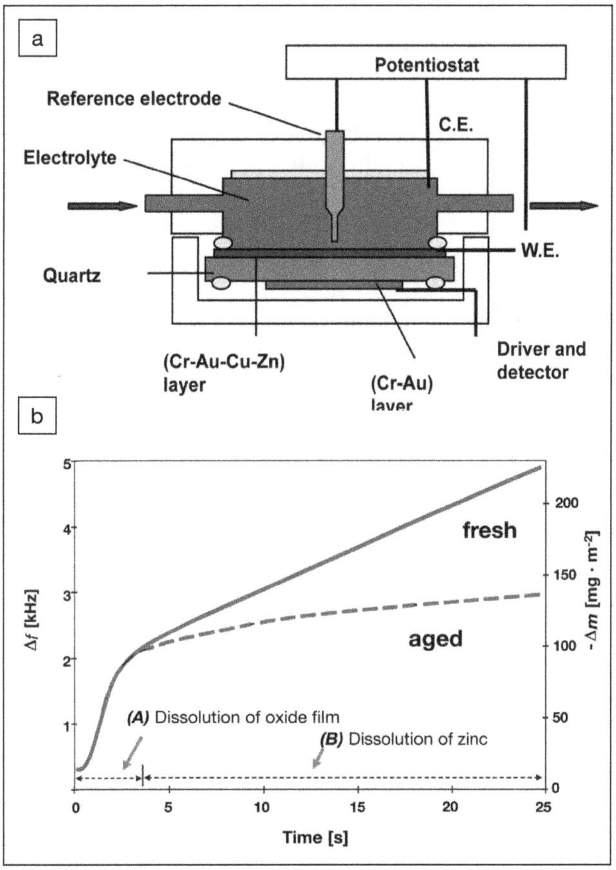

Figure 3.3: Investigation of cleaning reactions in alkaline solution by the Electrochemical Quartz Microbalance (ECQM). a) Setup of experiment; C.E. = counter-electrode, W.E. = working electrode; b) Oscillation frequency and mass changes over time and influence of bath age in decelerating the metal dissolution. Source: F. M. Androsch et al. [9]

Bath control usually involves the volumetric analysis of the alkaline content by titration, i.e. addition of an acid of known concentration until the alkali is neutralised, and a certain pH value is obtained. Indication can be made by pH measurement or by use of

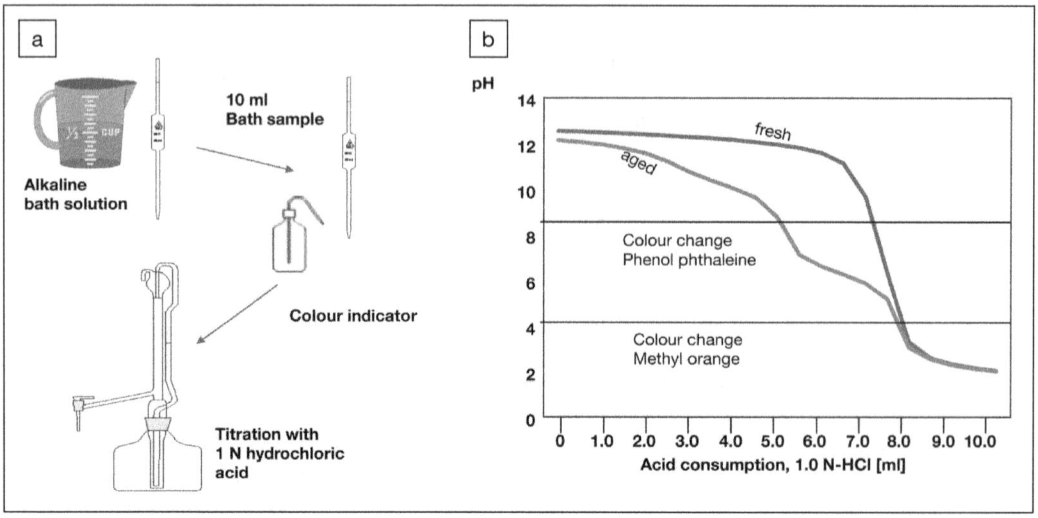

Figure 3.4: Bath control of alkaline cleaner baths by titration; a) procedure for titrimetric control; b) typical titration curves for fresh and aged bath solutions

an appropriate colour indicator. The titration curves show the effect of ageing by e.g. carbonation. Figure 3.4 illustrates the principle of this control method.

In industrial lines, an automated control of the cleaner bath is usually done by measurement of the electrolyte conductivity. For calibration of this method, however, the titration is still indispensable.

3.3.4 *Surfactants* [10]

Some consideration should be placed on surfactants as important factors in aqueous cleaning. These act through their amphiphilic nature that is caused by the molecular structure, having a polar moiety bonded to an unpolar, often longer hydrocarbon chain. Depending on the type of polar functionality, surfactants are classified as anionic, cationic or non-ionic. Examples are given in Figure 3.5.

Important anionic surfactant groups are the salts of fatty acids (soaps), sulphuric acid semi-esters of long-chain fatty alcohols or alkyl benzene sulphonates.

Anionic surfactants	
$C_{17}H_{35}-(CO_2)^-$	carboxylate (soap)
$H-(CH_2CH[CH_3])_4-C_6H_4-(SO_3)^-$	tetrapropylene benzene sulphonate (TPBS)
Cationic surfactants	
$C_nH_{2n+1}-NH^+([C_2H_4O]_mH)_2$	ethoxylated ammonium salt
$C_nH_{2n+1}-(CO)-N(C_2H_4-NR_3^+)_2$	quaternary ammonium base
Non-ionic surfactants	
$C_nH_{2n+1}-(CO)-O(C_2H_4O)_mH$	ethoxylated ester
$C_nH_{2n+1}-O(C_2H_4O)_mH$	ethoxylated fatty alcohol

Figure 3.5: Examples for the classification of surfactants re. functional groups controlling their bipolarity

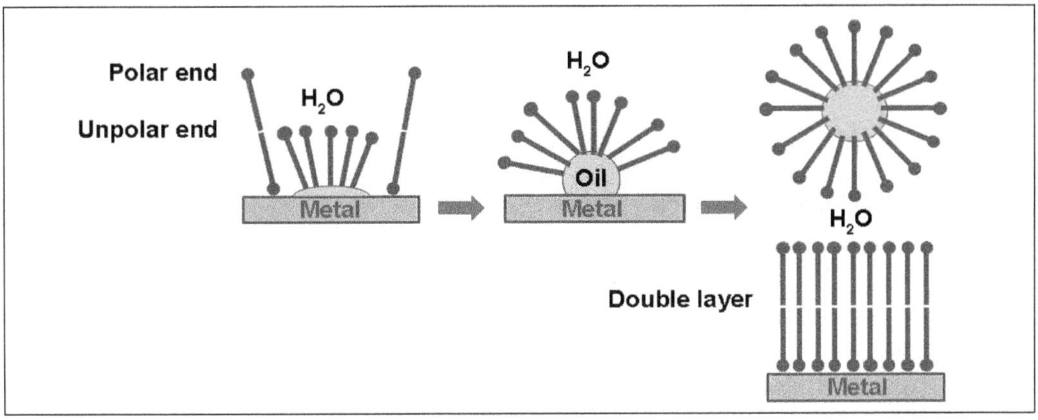

Figure 3.6: Removal of an oil drop from a metal surface by spontaneous alignment of surfactant molecules

Cationic species are e.g. quaternary ammonium bases that contain alkyl groups.

Typical representatives of the non-ionic classes are polyethoxylated alcohols or esters.

Surfactants assemble at the interface of non-miscible phases, self-aligning according to the polarity of the two phases and to their own dipolar character. They can displace oily matter (and any solid that is entrapped in it) from a metal surface. Figure 3.6 depicts the removal of an oil droplet by surfactant action.

The removed dirt particles are also reliably kept away from the substrate surface. Hence, a working cleaner solution accumulates dirt and oily particles which therefore have to be displaced continuously, to extend the useful service life of the bath. Surfactants are used in appropriate blends that enable the dirty emulsion to be split in a controlled way in separate tanks. The emulsion stability is adjusted according to the way of application of the cleaner by either spraying or immersion. Contaminants can be continuously removed from the bath and discarded. The surfactant mix must also effectively control the foam formation, particularly in the case of spray cleaners.

Figure 3.7: Foam generation in open waterways due to the accumulation of non-degradable surfactants Source: Wikipedia

Being surface active substances, surfactants interfere with biological reactions taking place at cell membranes, and hence are toxic to aquatic species. The use of surfactants is therefore subject to environmental regulations, to avoid the build-up of surfactants and their metabolic successors in the environment (cf. Figure 3.7), and to prevent harm to aerobic bacteria in public sewage works. By European and national law, surfactants used in detergents require to be readily biodegradable [11-13]. Accepted tests are codified by EU Directives and Regulations, and related ISO norms [14, 15].

3.4 Cleaning physics

3.4.1 Influence factors

As with any chemical reaction, the action of cleaners is substantially affected by physical factors [16]. The speed of cleaning depends on the concentration of the cleaning agent and the temperature. Typically, the speed is a linear function of the concentration, and a logarithmic function of temperature, so that 10 K increase result in a duplication of speed.

Since cleaning takes place on a solid surface, also the availability of sufficient amounts of the cleaner ingredients at any time becomes important. The cleaning action is decelerated when there is no relative motion between the work piece and the solution. Spray cleaners bring on sufficient motion in the reaction zone, as the spray jet impacts on the metal strip. Therefore, in principle, spray cleaners can be used at lower concentrations. Immersion cleaners require extra measures to ensure turbulent conditions at the surface. The speed of the strip in coil immersion lines may be supported by underwater injection nozzles.

Typical conditions for spray cleaning are cleaner concentrations of 0.5 to 1 %, pressures of 1 to 2 bar and temperatures of 50 to 70 °C that are applied during a period of usually 5 to 20 s.

Immersion cleaners are used at higher concentrations, and so-called low-temperature cleaners have not proven sufficient for the fast coil lines.

3.4.2 Rinsing and maintenance [17, 18]

Following the cleaning stage, excess cleaner solution, soil and reaction products are removed from the metal surface in a multi-stage rinsing section. The European Reference Documents on Best-available Techniques (BREF) require these rinse stages to be operated in a counter-flow cascade to achieve a most efficient use of water. The efficiency of the rinse increases exponentially with the number of rinse baths. The final rinse is usually fed with deionised

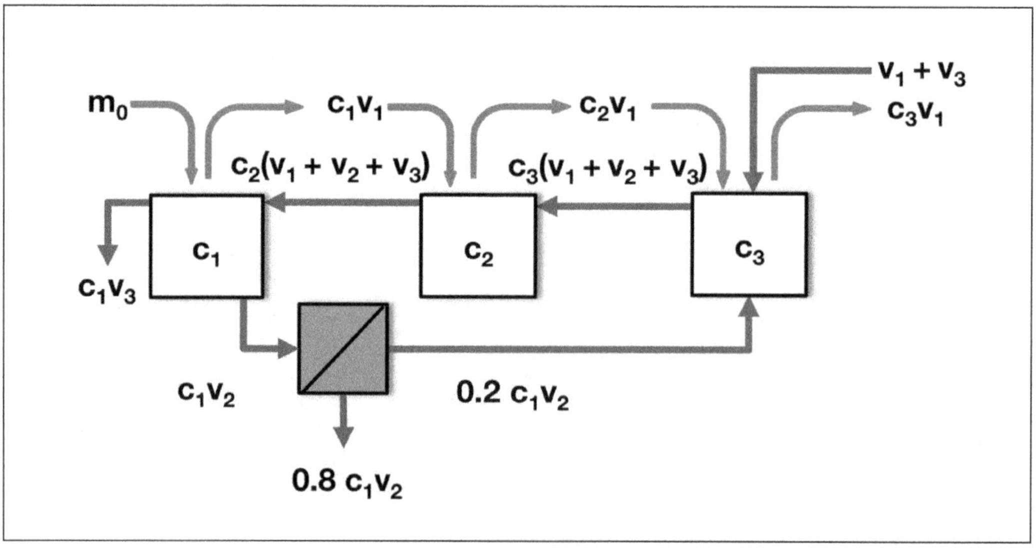

Figure 3.8: Schematic view of a rinse cascade system

water (DI). Such a DI rinse is mandatory when a so-called no-rinse pretreatment is used, to avoid high electrolyte concentration and contamination by unwanted, potentially corrosive ions. Usually a maximum acceptable electrolyte concentration is specified for the last rinse, given in units of the electrical conductivity. Recommended values range between 30 and 100 µS/cm at room temperature.

To minimise drag-out, the liquid film is generally removed in between individual stages, the drip-off supported by squeeze rollers etc. The loss of liquid can thus be reduced to 2 to 5 ml/m^2 with properly maintained squeeze rollers. Nonetheless, a controlled overflow is used to maintain a low level of contamination, and extend the bath service life. Continuous discarding and replenishing enables substantially longer service lives. An overflow system also results in lower water demand, and avoids extended downtime and cost involved with discontinuous dumping and make-up of the bath.

The counter-flow rinse cascade (see a schematic of the principle in Figure 3.8) is designed as required by the ratio between the amount of substances dragged in and the counter-flowing water volume (stationary state) according to BREF. The first rinse water is continuously deoiled (by demulsification, skimming, and ultra-filtration) and reused to replenish the main bath, together with an appropriate builder/surfactant package (cf. Chapter 3.3.6).

3.5 Literature

[1] Sander, J., Praktische Fragestellungen beim Einsatz technischer Reinigungsmittel auf Metalloberflächen (Practical Considerations Regarding the Use of Industrial Cleaners ...), Seminar Proc., Techn. Akad. Wuppertal, 1988

[2] Wichelhaus, W., Buetfering, L., Reinigung und chemische Vorbehandlung von Metallen (Cleaning and Chemical Pretreatment ...), Galvanotechnik 103, 2005, pp. 2712 ff

[3] Sander, J., Kirmaier, L., Manea, M., Shchukin, D., Skorb, E., Anticorrosive Coatings, Vincentz Network, Hannover 2010, pp. 23 f

[4] Asthana, R., Sobczak, N., Wettability, Spreading, and Interfacial Phenomena in High-Temperature Coatings, J. Metals e-Version, 52 (1), 2000, www.tms.org/pubs/Journals/JOM/0001/Asthana/Asthana-0001.html (19-06-13; 09:14 h)

[5] anon., Measurement of surface tension: Test ink and contact angle method, Press Release, Plasmatreat GmbH, Steinhagen 2007; www.plasmatreat.co.uk/measuring_surface_tension_determination.html (19-06-13; 09:17 h)

[6] Gerstenberg, K.W., Netzung, Oberflächenenergie und Young'sche Gleichung (Wetting, Surface Energy and Young's Equation), TIGRES Publication 01, Dr. Gerstenberg GmbH, Rellingen 2006, pp. 5 ff; http://www.tigres-plasma.de/de/publikationen/netzung-oberflaechenenergie-youngs-gleichung.html (19-06-13; 14:49 h)

[7] Sander, J. et al. [3], pp. 110 f

[8] Sander, J. et al. [3], pp. 27 ff

[9] Androsch, F. M., Stellnberger, K.-H., Wolpers, M., Jandel, L., Drescher, D., Sander, J., Seidel, R., Chromate-free Coil Coating and One Year of Production Experience, ECCA General Meeting, Proc., European Coil Coating Association, Brussels 1999

[10] Sander, J. et al. [3], pp. 30 ff

[11] Karsa, D. R., Porter, M. R. (eds.), Biodegradability of surfactants, Blackie, Glasgow 1995, pp. 1946 ff

[12] anon., BGBl. (Fed. Law Gazette) I (25) Art. 1.2, Bundesanzeiger, Bonn 1986, p. 851

[13] Karsa, D. R., Porter, M.R. [11], p. 1954

[14] anon., Regulation (EC) No 1272/2008 of the European Parliament and of the Council of 16 December 2008 on classification, labelling and packaging of substances and mixtures, amending and repealing Directives 67/548/EEC and 1999/45/EC, and amending Regulation (EC) No 1907/2006, Chapter 4.1.2.9. Rapid degradability of organic substances, Official Journal of the European Union L 353, 2008, pp. 1 ff, www.eur-lex.europa.eu/...

[15] anon., 2007/506/EC: Commission Decision 21 June 2007 establishing the ecological criteria for the award sof the Community eco-label to soaps, shampoos and hair conditioners (notified under document number C[2007] 3127, Appendix "Environmental Criteria", Part 3.a), Aerobic Biodegradability of Surfactants, Official Journal of the European Union L 186, 2007, pp. 36 ff, www.eur-lex.europa.eu/...

[16] Sander, J. et al. [3], pp. 32 ff

[17] Fresner, J., Zero Emission Retrofitting Method for Existing Galvanising Plants (ZERMEG), Austrian Fed. Min. Traffic, Innovation, Technology (ed.), Vienna 2003, p. 45; www.nachhaltigwirtschaften.at/nw_pdf/0321_zermeg.pdf (19-06-13; 15:16 h)

[18] Bosse, K. (ed.), Entwurf des deutschen Beitrags zu den besten verfügbaren Techniken bei der Behandlung metallischer und nichtmetallischer Oberflächen mit chemischen und elektrochemischen Verfahren (Draft Best Available Techniques for the Treatment of Metallic … Surfaces with Chemical … Processes), AG BREF Oberflächentechnik, Umweltbundesamt (Fed. Environment Agency), Berlin 2003, pp. 49 ff; www.bvt.umweltbundesamt.de/archiv/oberflaechenbehandlungvonmetallen.pdf (19-06-13; 15:21 h)

4 Pretreatment

4.1 Introduction

Native oxide layers may serve well as efficient anti-corrosive barriers for bare bulk metal, but from the perspective of technical surfaces, it is often desired to create a more uniform, controlled passivation layer, to better preserve the surface aspect and features. To this end, it is necessary to remove the native oxide film, and replace it by a similar film under controlled conditions, with improved features, for instance, uniform thickness, lower porosity, better transparency, or higher electrical resistance. The passivation is effected by formation of a mechanical and/or electronic barrier against corrosive substances and media. Furthermore, underneath an organic coating, it is important to provide a good adhesion to that coating. All this is achieved by means of a conversion pretreatment. Usually, a conversion coating is specifically adapted to the substrate metal. Conversion coatings usually have specific weights between 50 and 500 mg/m^2 and gauges of several 10 nm. A good conversion pretreatment has to form chemically bonded layers that are insoluble in water under changing pH conditions. A good conversion coating must also provide long-term corrosion resistance upon mechanical damage [1, 2].

Features and tasks of pretreatments are:

- Covering of the clean, reactive surface with a thin, non-metallic coating, with typical specific mass 50 to 500 mg/m^2, gauge 20 to 100 nm
- Most processes with chemistry adapted to substrate
- Passivation by formation of a mechanical and/or electronic barrier against corrosive substances and media
- Adhesion promotion towards paint coat
- High resistance under all environmental conditions
- Corrosion resistance and resistance against creepage under organic coatings after mechanical damage

4.2 Substrates

4.2.1 Overview: Particular substrates for coil coating

A variety of materials are used for metal construction, of which most surfaces are coated with organic paints. The most important of these metals are (carbon) steel and aluminium. In many cases, steel is protected against corrosion with an additional zinc coating. Obviously, in coil coating, the substrate is a flat, rolled product with superior formability properties, and a very high dimensional accuracy and surface quality. Tables 4.1 and 4.2

Table 4.1: Important substrates for coil coating: Features and major uses

Substrates				Use		
Base metal	Coating	Gauge in [µm]	Code	Architectural	Appliance	Transport
Cold-rolled	none		Steel (CRS)	indoor		
	electrolyt. galvanised	3–7.5	ZE, EG		x	x
	hot-dip galvanised	7–30	Z, HDG	in and outdoor	x	x
	Galfan	7–30	ZA	in and outdoor	x	x
	Galvalume	15–30	AZ	outdoor		
Aluminium				in and outdoor	x	x

Table 4.2: Metal substrates in industrial painting

Ferrous metals	Non-ferrous metals
Steel, stainless steel	Wrought & rolled aluminium (mainly Mn and Mg alloys)
Metallic-coated steel	Magnesium
- Galvanised, alloy-galvanised zinc	Zinc sheet
- Electro-galvanised, alloy-galvanised zinc	Others (copper, brass, etc.)
- Aluminised	
- Tinplate	

summarise important substrates for industrial painting in general, and in more detail the substrates that prevail in coil coating with an indication of their primary uses [3].

4.2.2 Steel

4.2.2.1 Blast furnace process

As the most common construction metal, steels of ever-increasing quality and an abundance of specialised compositions and preparations have been in use for many centuries. The worldwide production of crude steel in 2012 amounted to 1,550 Mt. 10.9 % of this tonnage were generated in the EU-27. The basic constituent of steel, iron, is produced in blast furnaces (Figure 4.1a), by smelting the ore with coke and lime under an oxygen-depleted atmosphere. The molten metal is obtained from the run-off of the furnace (Figure 4.1b), and separated from the silicate/carbonate slag floating on the melt [4-9].

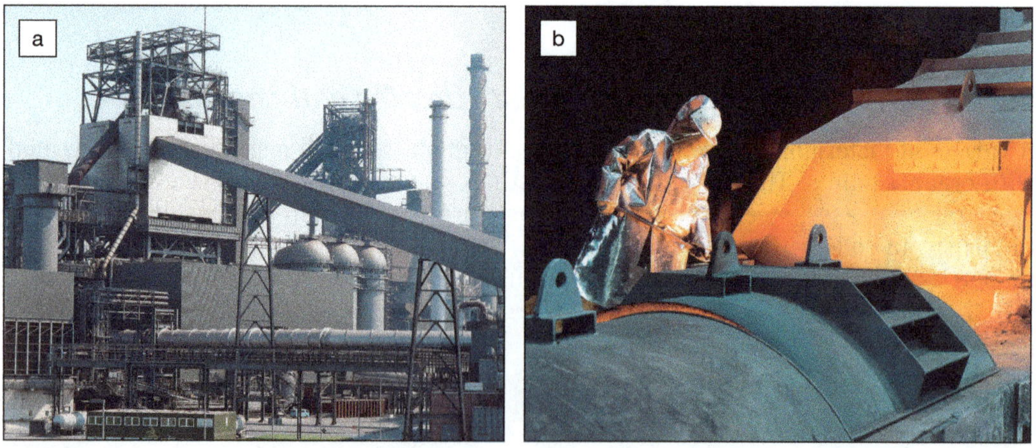

Figure 4.1: a) Modern blast furnaces; b) tapping of molten iron

Source: ThyssenKrupp Steel Europe

Figure 4.2: a) Molten iron is poured into the converter for steel making; b) large amounts of steel scrap are recycled, providing a source for alloying elements and oxygen Source: a) ThyssenKrupp Steel Europe; b) ThyssenKrupp Rasselstein

4.2.2.2 Steel making and refining

The crude iron melt is then converted to steel by controlled oxidation. Further alloying and annealing processes are used to produce a wide range of steels of different quality (cf. Figures 4.2a and b). The choice of steel is made with respect to the intended use. Particular qualities and strength properties are required for roll profiling, bending, deep-drawing and other metalworking operations. Highly alloyed steels display particular qualities, like stainless steel (alloying elements: chromium, manganese, nickel, molybdenum). However, only minor amounts of these materials are subject to painting operations.

4.2.2.3 Galvanised steel

Sheet steel of high surface quality is produced by cold rolling, further reducing the gauge of hot-rolled feedstock. Cold-rolled steel sheet is used for indoor applications like furniture, some appliances or lighting.

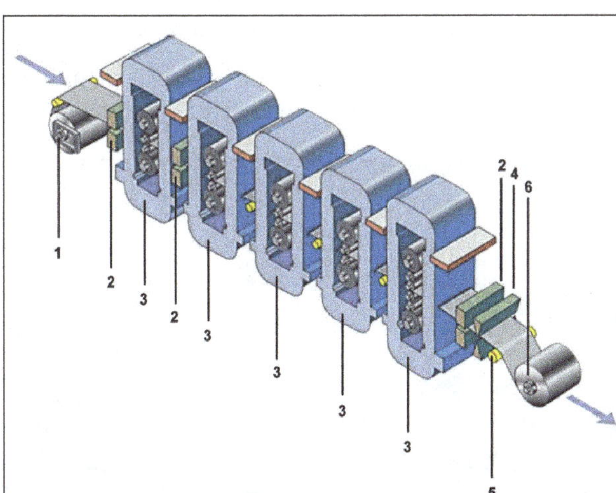

Figure 4.3: Schematic representation of a cold rolling mill; 1 = Pay-off reel; 2 = Gauge control; 3 = Rolling stands (4 x 4-high, 1 x 6-high); 4 = Cross-cut shear; 5 = Flatness control; 6 = Tension reel (recoiler) and strap winder Source: Salzgitter Stahl

However, low-alloyed carbon steel, when exposed to the atmosphere, quickly develops a layer of oxide scale, composed of FeO (wuestite), mixed oxide Fe_3O_4 (magnetite) and Fe_2O_3 (haematite), on its surface. Therefore, when intended for outdoor use, e.g. car manufacture, façade cladding or roofing, most of the material is covered with a

Table 4.3: ZnMgAl coated substrates in coil coating Source: Schulz, J.[12]

Steel company	Arcelor Mittal	Salzgitter	Tata	ThyssenKrupp Steel Europe	Voest
Trademark	Magnelis	Folastal Strongcoat	"Colorcoat" based on "Magizinc"	"Pladur" based on ZM "EcoProtect"	"Colofer" corrender coated
Certificate*		DIBt Z-30.11-42 18.12.2009	DIBt Z-30.11.36 (11.11.2009 / 08.02.2010)	DIBt Z-30.11-30 (18.02.2008) CSTB Evaluation Technique 2/09-1389 (01.06.2010)	DIBt Z-30.11-37 (20.01.2010)
Organic coating [µm]		SP 15 SP 25	SP 15 SP 25	SP 25 PVdF 25	SP PUR PVdF
Minimum coating [g/m²]		100 (w. SP 15) 140 (w. SP 25)	100 (w. SP 15) 140 (w. SP 25)	130	120
Alloying elements [%]	Mg: 3.0 Al: 3.7	1.6 1.6	1.6 1.6	1.0 1.0	2.0 2.0

* *DIBt = Deutsches Institut für Bautechnik (German Inst. f. Building Technology);*
CSTB = Centre Scientifique et Technique du Bâtiment (French Scientific and Techn. Centre of Building)

zinc (Zn) layer by hot-dip or electro-galvanising, in order to prevent the reddish-brown iron oxide formation (rust).

Zinc coatings serve as a first protective barrier against corrosion. Due to its lower electrochemical potential zinc is less noble than iron, and therefore corrodes preferentially. However, this corrosion reaction is kinetically retarded by formation of a dense surface layer of mixed zinc oxide, hydroxide and carbonate. Moreover, these corrosion products are colourless (white rust). The mechanical and technological specifications of both the base steel and any metallic coatings are subject to industrial standards.

On hot-dip galvanised sheet metal, a zinc coating is produced in fast, continuous coil lines, immersing cleaned, annealed steel strip in a zinc melt. The zinc melt may contain aluminium (Al), as well as small amounts of other alloying elements, e.g. silicon (Si), lead or antimony. Specified alloy-galvanised steel sheet is known as "Galfan", ZA (5 % Al), or "Galvalume", AZ (55 % Al, 1.6 % Si). After the zinc melt, the strip is cooled down by pressurised air (air knife, cf. Figure 4.4) in order to control the Zn thickness. Nowadays, a spangle-free, finely crystalline surface is state of the art.

Figure 4.4: Cooling and adjusting the thickness of the zinc coating in a high-capacity hot-dip galvanising line Source Salzgitter Stahl

More recently, ternary systems with low amounts (<4 % each) of magnesium and aluminium have entered the marketplace [10-12]. Several commercial products have been invented that differentiate in the composition (alloying grade) and the gauge of the galvanised layer. These metallic coatings have been shown to provide a corrosion resistance superior to Galfan coatings, both in the bare state and coil coated. This performance level is achieved even with only 130 g/m^2 (approx. 10 μm) of the new metallic coating, i.e. about half the thickness of an ordinary galvanised layer. In particular, corrosion at cut edges is substantially reduced.

Electro-galvanised sheet is produced in a continuous coil operation by electro-deposition of Zn or Zn alloys from an electrolyte solution onto the clean, annealed and temper-rolled steel strip. The Zn layer gauge can be adjusted by choice of the parameters of the electrolysis.

4.2.3 Aluminium

4.2.3.1 Bauxite refining and Al electrolysis

Even less noble than zinc, Al [13-17] is found almost exclusively in its oxidised state in silicate and carbonate ore, i.e. clay and bauxite minerals. Bauxite contains approx. 50 % Al oxides. Being discovered in the early 19th century, elementary Al took five decades before being produced on an industrial scale. Today's improved process involves hot extraction of Al oxide from the mineral with sodium hydroxide solution, from which aluminium hydroxide, gibbsite, $Al(OH)_3$, is precipitated upon cooling, and filtered off.

Dehydration at 1,100 °C in fluidised bed furnaces renders pure aluminium oxide, Al_2O_3. Elemental Al is derived from the oxide by electrolysis in molten cryolite, $Na_3(AlF_6)$ (Figure 4.5a). The process requires a high amount of energy, i.e. up to 16 kWh per kilogramme of Al. Refining and alloying of the aluminium is done by blending with appropriate amounts of alloying elements and aluminium scrap. As the recycling of Al scrap takes much less energy than the primary production, the material that is recovered (secondary Al) is an important source for the aluminium industry. In 2007, the world production was 38 Mt of primary and 9.6 Mt of secondary Al. Finally, the refined aluminium is cast into manageable forms for further processing. Stock for aluminium flat products is often cast into large slabs (Figure 4.5b). Also belt or strip casting is used to produce feedstock that is closer to the end dimensions.

Subsequent hot and cold rolling processes are employed to reduce the thickness of the ingot to the finally required metal gauge. The typical thickness of Al sheet metal in coil coating ranges from 1.5 down to 0.4 mm. Strip casting may allow the upper end of this gauge range to be achieved in a single pass in a tandem hot mill. Lower thicknesses are obtained

Figure 4.5: a) Electrolysis cells in an aluminium plant; b) aluminium slabs in storage prior to rolling

Source: a) Norsk Hydro; b) Rio Tinto Alcan

Table 4.4: Composition, properties and uses of aluminium alloys

Alloy	AA*-code	Properties	Uses
Al 99 % min	1xxx		Foil, sheet, lithographic sheet
Cu	2xxx	+ mechanical – corr. prot.	Aerospace industry
Mn	3xxx	+ mechanical ~ deep drawing	Food packaging architecture, heat exchangers
Si	4xxx	+ casting	Gen. construction, heat exchange
Mg	5xxx	+ fastness after cold forming	Food packaging car, transport, architecture
Mg, Si	6xxx	+ fastness	Car, architecture, transport
Zn	7xxx	– corr. prot.	Aerospace, heat exchangers
others	8xxx		Foil (Fe), aerospace industry (Li)

* *The Aluminum Association Inc.*

by cold rolling. Al semis for food containers and beverage cans and lids are produced in gauges down to 150 μm. Even thinner material is used for flexible packaging (cf. Chapter 9).

Thermal treatment by controlled annealing and cooling, as well as tempering render the material with the required final technical features and surface conditions. Figures 4.6a and b show a reversing breakdown mill for the initial gauge reduction and the end product of the rolling process, Al coils of high surface quality that are ready for use in coil coating.

4.2.3.2 *Properties of aluminium and its alloys*

Despite its low standard potential, technical aluminium is resistant to corrosion, since it forms dense surface oxide layers under normal atmospheric conditions. These layers inhibit ongoing structural corrosion, as long as the metal surface is neutral and no contact corrosion with more noble metals is possible. Hence, corrosion phenomena on Al surfaces are prominent with alloys that contain copper or when acidulated conditions apply, e.g. with the so-called filiform corrosion. A large variety of Al alloys are used for different technical purposes.

Table 4.4 shows typical alloys and their applications. In particular, cold rolled wrought Al alloys are common substrates for coil coating, i.e. especially the 1000, 3000 and 5000 series. For automotive sheet, also AA 6000 alloys are used.

Figure 4.6: a) Aluminium reversing rolling mill; b) cold rolled aluminium coils *Source: a) Novelis; b) Norsk Hydro*

4.3 Basics of pretreatment technology

Adhesion of a coating is highly dependent on the type of substrate and its proper cleaning and conversion treatment (pretreatment) by an adjusted process, that also provides an improved corrosion resistance and durability. The painting operation, therefore, includes preceding process steps for cleaning and pretreatment. Most conversion coatings are applied in spray or immersion processes. In particular in modern coil coating lines, the chemical solution is often adjusted to a no-rinse process, i.e. applied at an appropriate dilution onto the clean, dry strip surface where it immediately reacts off to form transparent, uniform layers upon drying at temperatures above 50 °C. Neither chemicals in excess nor products of side reactions are left, hence no rinsing is required after pretreatment.

Historically, alkaline passivation and chromating processes have been widely used. These processes require a fresh water rinse to remove excess chemicals and by-products. Mostly, chromate post-rinses have been used to enhance the corrosion resistance. However, chromate-free processes are favoured nowadays, as they avoid the use of the highly toxic and hazardous substances. Cr-free chemistry is mandatory with automotive and electric equipment.

After any pretreatment, the work piece surface must be completely dry before being processed to the paint section. Indirectly fired hot-air, infrared or induction dryers are state of the art.

4.4 Pretreatments for coil coating

4.4.1 Alkaline passivation

Alkaline passivation [18] has been the predominant pretreatment process for hot-dip galvanised coil. The treatment solution comprises caustic soda, iron (Fe) and cobalt (Co) compounds and solubilising complexants. Optionally nickel ions (Ni) could be present. In the course of the reaction, a zinc oxide coating is formed that incorporates the Fe and Co (or Ni) ions in a spinel-type lattice. The alkaline passivation exhibits outstanding adhesion, thus allowing extreme deformation, however it requires a post-rinse to obtain sufficient corrosion protection. Typical specific coating weights range between 0.2 and 0.4 g/m^2.

The conversion layer is completed by a post-rinse that contains chromium (III) and chromate ions. Alkaline passivation is suitable for all galvanised substrates, including the aluminium alloyed, Galfan and Galvalume. More recently, the use of heavy metals like Cr, Ni, that form toxic compounds, and in their wake, Co, has become a matter of concern. As a result, chromium-free post-rinses have been introduced. In Europe, some development efforts on the alkaline passivation process have been directed towards the replacement of the heavy metals by magnesium (Mg), however these developments have not taken ground.

Equation 4.1: Cobalt oxidation

$$2\ Co(OH)_2 + 0.5\ O_2 \rightarrow CoO(OH) + H_2O$$

Equation 4.2: Generation of a zinc-cobalt spinel

$$2\ CoO(OH) + Zn(OH)_2 \rightarrow Zn(Co_2O_4) \downarrow + 2\ H_2O$$

Equation 4.3: Generation of a cobalt-iron spinel

$$2\ Fe^{3+} + Co^{2+} + 8\ OH^- \rightarrow Co(Fe_2O_4) \downarrow + 4\ H_2O$$

Key reactions of the film formation during alkaline passivation are represented by the above Equations 4.1 to 4.3. These describe the oxidation of a part of the Co from the divalent to the

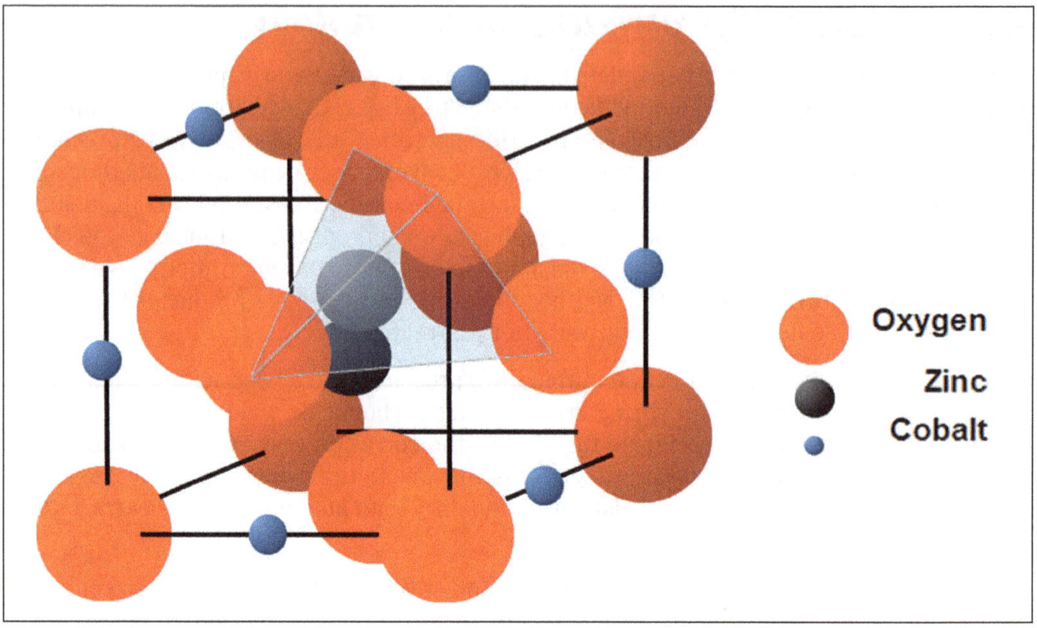

Figure 4.7: Spinel structure of $Zn(Co_2O_4)$

trivalent state, and the precipitation of the Zn, Co and Co, Fe spinels. The crystal structure of these spinel-type mixed oxides is illustrated in Figure 4.7.

4.4.2 Iron phosphating

The term "phosphating" is used for two different classes of pretreatment, i.e. iron phosphating [19], and zinc phosphating. However, the latter, i.e. the precipitation of crystalline zinc phosphate conversion coatings on the metal, does not play an important role in coil coating in the European market.

Iron phosphating comprises the treatment of steel surfaces with acidic sodium phosphate solution (dihydrogen phosphate). Iron is dissolved by the attack of the acid, and a thin, amorphous iron phosphate layer is formed, as laid out in Equations 4.4 and 4.5. The thickness of the conversion coating can be increased by the use of oxidants like molybdate, chlorate or organic nitro compounds that inhibit the formation of hydrogen, thereby accelerating the net reaction.

Equation 4.4: $6\ Fe + 3\ NaClO_3 + 18\ H^+ \rightarrow 6\ Fe^{3+} + 3NaCl + 9\ H_2O$

Equation 4.5: $Fe^{3+} + H_2PO_4^- \rightarrow FePO_4 \downarrow + 2\ H^+$

On steel coil, the process is operated in spray lines. The conversion coating ranges at specific weights between 0.2 and 0.5 g/m^2. Iron phosphating renders a basic level of corrosion protection. To enhance the performance, the final rinse is carried out with a sealant, including chromate chemicals.

4.4.3 Chromating

4.4.3.1 Yellow chromating (chromium chromating)

Historically, the unique properties of chromium have been the key to most versatile corrosion protection uses. Most pretreatments have comprised at least one processing step with chromate containing chemicals to obtain the necessary corrosion resistance of the final product. The very high oxidative power of the chromate ion in acid systems results in a very fast superficial redox reaction on any metal. It forms an insoluble, very dense mixed oxide layer of Cr^{3+} and the substrate metal, e.g. Zn. The cathodic branch of this reaction has a standard potential of E_{pot} = 1.195 V. The surface reaction is characterised by acidic attack of zinc and subsequent film formation by partial reduction of chromic acid. Owing to these reactions, the acid is consumed and zinc dissolves.

The formation of such a conversion film is described in Equations 4.6 to 4.8. Similar considerations apply for cold-rolled steel or aluminium substrates [20, 21].

Precipitation of zinc/chromium chromate (Yellow chromating):

Equation 4.6: $Zn + 2\,H^+ \rightarrow Zn^{2+} + 2\,[H]$

Equation 4.7: $Zn^{2+} + (CrO_4)^{2-} \rightarrow Zn(CrO_4)\downarrow$

Equation 4.8: $2\,(CrO_4)^{2-} + 3\,[H] + 4\,H^+ \rightarrow Cr(OH)(CrO_4)\downarrow + 3\,H_2O$

The long-lasting corrosion resistance is also due to the so-called self-healing effect. This is usually attributed to residual, mobile chromate ions that are incorporated in the oxide layer. In case of a mechanical damage of the passivation layer that penetrates through the underlying substrate metal these chromate ions are available for a quick passivation reaction at the newly exposed metallic location.

Single-stage spray chromating has been used on Al substrate. On galvanised steel, two-stage processes with chromate-containing post-rinse have been common. Also chromate no-rinse products are available.

4.4.3.2 *Green chromating*

For aluminium, the so-called green chromate coating is popular. This name stems from the colour of the chromous (Cr^{3+}) phosphate layer that is formed by precipitation from a phosphoric acid containing chromate pretreatment solution. The formation of green chromous phosphate layer is depicted by Equations 4.9 and 4.10. While the chemical contains chromate, the coating is free of chromium (VI).

Precipitation of chromium phosphate (Green chromating):

Equation 4.9: $Al + 3\,H^+ \rightarrow Al^{3+} + 3\,[H]$

Equation 4.10: $(CrO_4)^{2-} + 3\,[H] + 3\,H^+ + H_2(PO_4)^- \rightarrow Cr(PO_4)\downarrow + 4\,H_2O$

Since chromous compounds are almost insoluble and non-toxic, such conversion coatings have been used for decades on Al surfaces, as a replacement for “yellow” chromium chromate treatments, and in particular as pretreatments in the food-related sector (canning, i.e. food containers, beverage cans and lids). Processes are available for immersion, spray, and no-rinse application.

However still, the immersion or spray products involve chromate that has been listed as a “Substance of Very High Concern” according to the REACH regulations in 2010. Following the removal from automotive and electrical equipment applications that essentially has taken place, driven by the respective directives, chromate chemicals are now likely to be banned altogether. An intensive search for chromate-free alternatives has been undertaken for quite a while.

4.4.4 [Ti,Zr]F_6^{2-} based processes: Chromium-free pretreatment [22-25]

4.4.4.1 Basics

In anticipation of a more stringent environmental protection policy, chromium-free pretreatment processes have been developed and introduced since the early 1990s. They are available for single work pieces (batch painting) e.g. window frame profiles and façade parts, as well as for coil coating.

In coil coating, chromium-free pretreatment chemicals based on aqueous solutions of titanium (Ti) and/or zirconium (Zr) fluoro complexes have been commercial since 15 years [22-25]. They are suitable for application with single-layer, high-durable polyester paints. Similar pretreatments have been industrially used since 1998, in particular in roll-coater applications, for virtually all relevant substrates in coil coating.

4.4.4.2 Chemistry

The Ti, Zr fluoro complex technology employs aqueous pretreatment solutions containing Ti, Zr complex fluorides, phosphoric acid, and water-soluble organic polymers. Optionally, other metal compounds may be present. The solutions are applied by a no-rinse process on to the clean strip and reacted and dried off at approx. 50 °C. In the course of reaction, an oxide layer is deposited with different metals (from the substrate and solution) and phosphorus in the cationic places of the oxide lattice. This inorganic layer is covered by a surface film deriving from the polymer. Chromium-free conversion coatings have gauges in the order of a few 10 nm, corresponding to Ti or Zr build-up values between 1 and 30 mg/m^2.

Pretreatments of this kind are used for a variety of substrates, such as aluminium, zinc, zinc alloys and cold-rolled steel. Also, the new ZnAlMg substrate is pretreated with unaltered processes that have been in use for state-of-the-art galvanised substrates. However, the physics and surface chemistry of the new substrate, i.e. lower flexibility and higher sensitivity to moisture, seem to require adaptations of the pretreatment chemicals. For certain uses, also molybdate compounds are added.

The key reactions during the formation of such chromium-free conversion coatings can be sketched as in the Equations 4.11 and 4.12. The hydrofluoric acid (HF) that is formulated as a by-product in the last equation is consumed by other reactions, e.g. formation of mixed Al oxyfluorides. In the net result, no free fluoride is released.

Equation 4.11: Dissolution of zinc

$$Zn + 2\,H^+ \rightarrow Zn^{2+} + H_2\uparrow$$

Equation 4.12: Precipitation of a hydrated mixed oxide/phosphate

$$(Ti,Zr)F_6^{2-} + H_3(PO_4) + Zn^{2+} + 3\,H_2O \rightarrow Zn(OH)_2 \cdot (Ti,Zr)(OH)(PO_4)\downarrow + 6\,HF$$

4.4.4.3 Conversion layer composition

Analysis of the conversion layer was reported using Auger and photoelectron spectroscopy combined with argon sputtering technique. Figure 4.8 shows the depth profile obtained by this method. Zinc, originating from the substrate, is found in all parts of the layer, albeit in higher concentration close to the interface between metal and pretreatment. This also applies to alloying elements of the galvanising layer, like aluminium. The transition metals (Me 1 and 2), and fluorine (F) that are ingredients of the pretreatment, are generally distributed throughout the layer, following the same pattern as oxygen (O). The concentra-

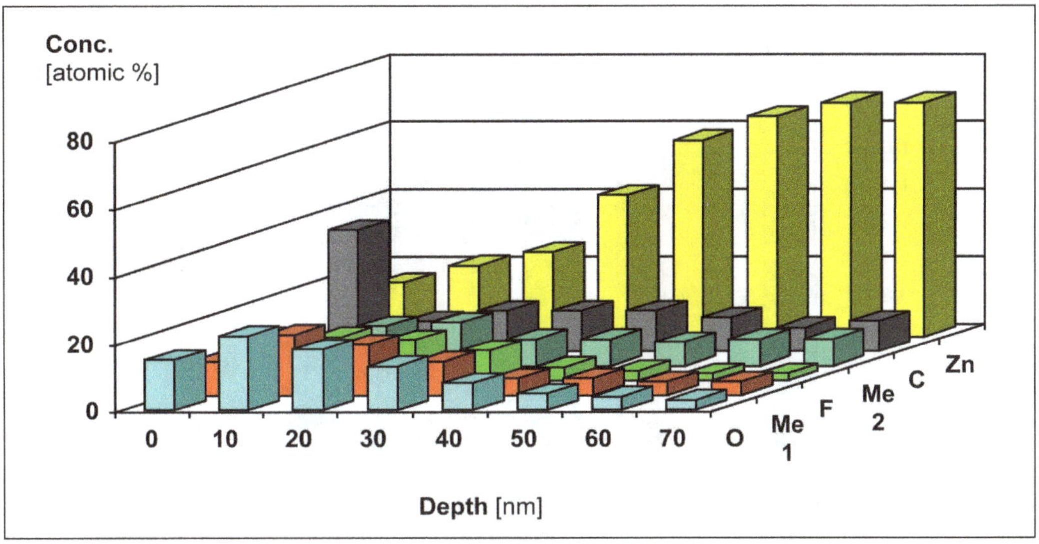

Figure 4.8: Elemental analysis across a chromium-free conversion coating by XPS/Ar sputtering

Source: Henkel

tion of carbon (C), indicating the presence of the organic polymer, increases sharply close to the outer surface.

The mechanisms of bonding and corrosion resistance are attributed to the metal oxide layer that forms in an initial reaction on top of the treated metal surface and then is covered by the organic matrix. Figure 4.9 shows a corresponding schematic. Thermal and calorimetric experiments, micrographic methods and spectroscopy of the layers are consistent with this picture.

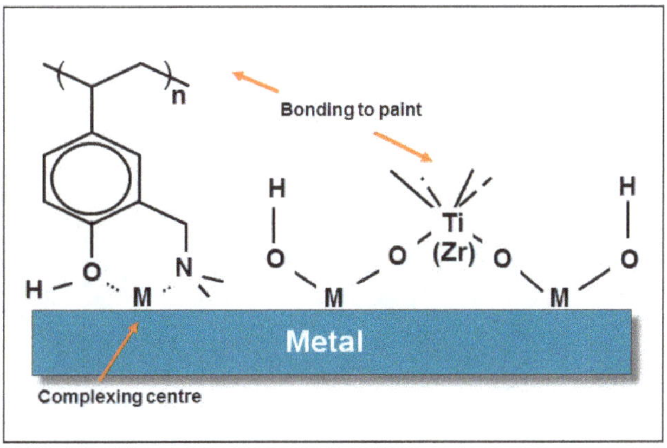

Figure 4.9: Schematic representation of the adhesion promoting components in a chromium-free conversion coating, i.e. a metal complexing, water-dispersible polymer (left) and a Ti or Zr oxo-complex (right).
M = substrate metal atom

Source: Henkel

4.4.4.4 *Performance*

The comparability between chromium-free and chromate pretreatments in various outdoor and accelerated corrosion tests has been shown in various papers. Figure 4.10 displays some laboratory results, showing the maximum creepage values from the scribe mark as were reported under commercial polyester coil coating systems in comparison to standard chromate spray and no-rinse pretreatments. In the same paper [26], salt spray, and cyclic corrosion test results including VDA and prohesion tests were shown to correlate well with 3 years outdoor exposure results from Hook of Holland in a statistical comparison. Moreover, the chromium-free pretreatment provides good process stability and consistent quality [27].

The anti-corrosive effectiveness of the Ti, Zr fluoro complex technology was demonstrated by a series of experiments using a special electrochemical method, the "height-regulated Scanning Kelvin Probe" technique (HR-SKP) [28]. The experiments and results are illustrated in Figure 4.11.

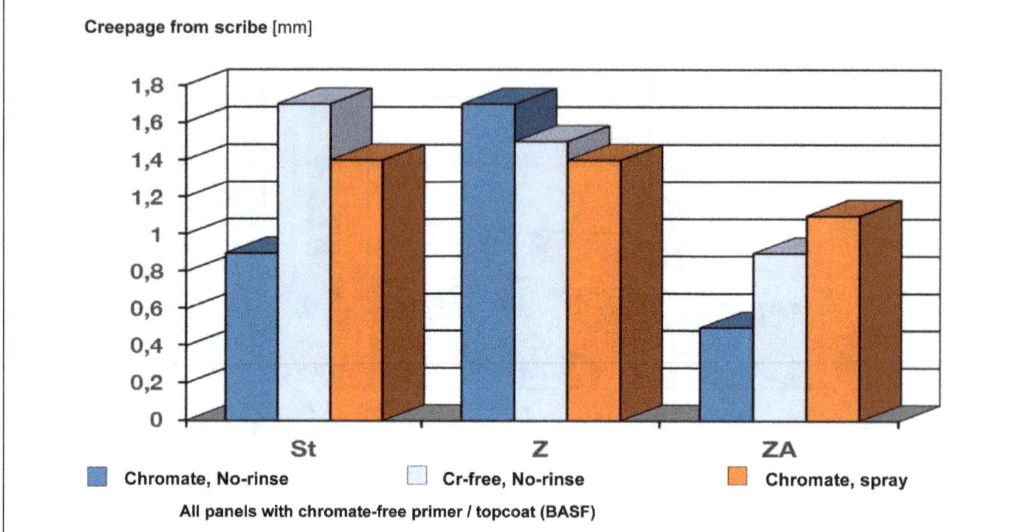

Figure 4.10: Anticorrosive performance of a chromium-free conversion coating after 1,000 h Neutral Salt Spray Test

Source: Henkel

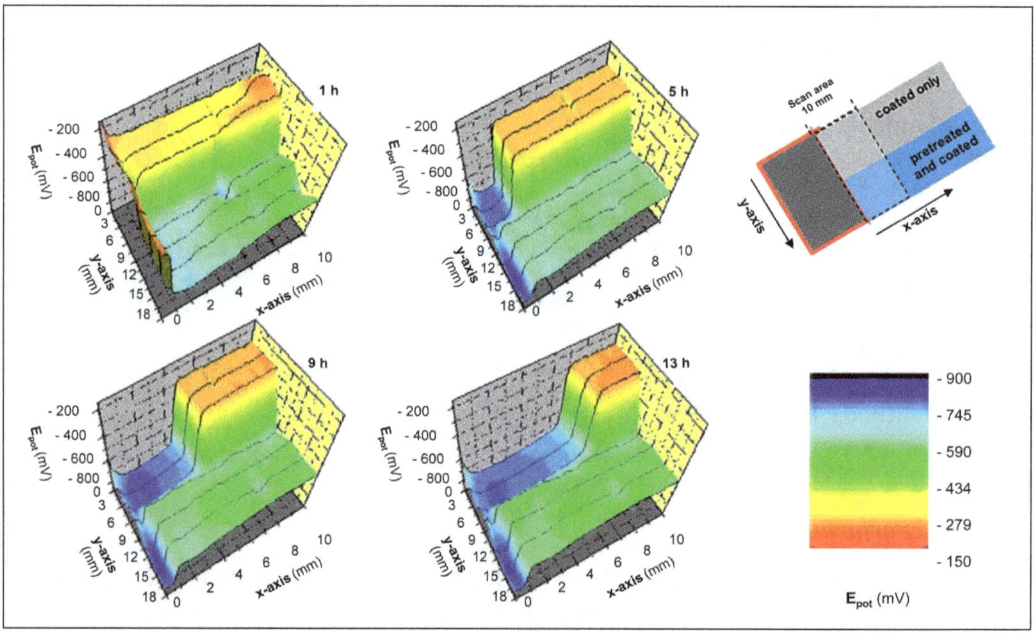

Figure 4.11: Model studies with the Scanning Kelvin Probe on a chromium-free conversion coating. Corrosion propagation is effectively stopped on the pretreated lower half of the test specimen

Source: Elsevier 2007; cf. Klimow, G., et al.[28]

Electro-galvanised panel samples were cleaned and etched following a standard laboratory procedure, covered with a protective tape across one end (forming the y-axis), then, perpendicular to it, along the x-axis, pretreated on half of the remaining bare surface. The entire specimen was then coated with a model clearcoat.

Upon removal of the protective tape, the specimen displayed a treated and an untreated area adjacent to each other, both coated with the clearcoat and having a sharp border to a

Figure 4.12: a) Chromium-free pretreated Al V-groove panel cladding on an industry façade; b) the dripping edge has no sign of corrosion after 10 years of service Source: Henkel

bare metal surface. One part of the bare surface (left area in the schematic) was encased with small stripes of resin, to form a small container which was filled with 0.5 M sodium chloride solution.

The HR-SKP device was used to scan the area 10 mm along the x-axis into the coating. It monitors the electrical potential underneath the paint film and its propagation from the defect with time, delivering a vivid picture of the advancing corrosion. It can be clearly seen that the corrosion is effectively withheld by the pretreatment, while on the unprotected surface, the drop of the potential and the corresponding delamination move on very fast. One possible explanation from the experiment is that, as the starting potential underneath the pretreated organic coating is very close to the Zn/Zn^{2+} standard potential (E_{pot} = -0.76 V) of the substrate, the overall corrosion reaction is not driven thermodynamically. Independent results converse with this view, suggesting a band gap widening in a Ti/Zr oxide layer that effectively suppresses oxygen reduction.

4.4.4.5 Architectural application

Chromium-free treated aluminium coil stock has been in use for over 10 years, for wall cladding and sun blinds. Figure 4.12 displays a part of a façade clad with coil-coated V-groove aluminium panels after 10 years of service life. The panels, along with sun blinds, prepared with a chromium-free no-rinse pretreatment and top-coated with high-durable polyester, are mounted on a south-facing wall, covering an area of about 1,200 m^2. The façade is exposed to sun, wind and rain in an industrial site, and also encountering mechanical damage around loading areas. It does not show any corrosion even at critical points like punch-holes, bolt and screw joints or dripping edges [29].

4.4.5 Synopsis: State-of-the-art coil coating pretreatments

Table 4.5 provides an overview on pretreatments in coil coating, including both conventional and alternative, chromium-free processes.

In particular, hot-dip galvanised (Z) strip still is commercially pretreated with alkaline passivation, while cold-rolled steel (St) used to be treated with iron phosphating solutions. All

Table 4.5: Pretreatments in coil coating - conventional and chromium-free replacement processes

Source: Meuthen, B., Jandel, A.-S. [8], pp. 53, 56

Process	Substrate[a]	Remarks
Alkaline passivation	Z, ZA, AZ, ZE	Chromate post-rinse Cr-free post-rinse
Yellow chromating	Z, ZA, AZ, ZE, Al	Chromate post-rinse[b]
Cr-free pretreatment	Z, ZA, Al	Spray and dip, optionally Cr-free post-rinse
Green chromating	Al	Food-safe Cr-free
No-rinse process	Z, ZA, AZ, ZE, Al, St	Chromate-containing Cr-free
Iron phosphating	St	Chromate post-rinse Cr-free post-rinse

[a] *Abbreviations: Z: hot-dip galvanised, ZA: Galfan, ZE: electro-galvanised, AZ: Galvalume (AluZinc), Al: Aluminium, St: cold-rolled steel*
[b] *except Aluminium*

of these pretreatments, if not chromate-based by themselves, are completed by chromate post-rinses.

The entire range of pretreatments for coil, however, can be replaced by chromate-free alternatives. Nevertheless, in the European coil coating market, such chromium-free pretreatments have only recently gained share, and have grown to approx. 5 % of the pretreatment use in 2009. Chromates have still prevailed in the largest target market, building applications, where very long warranty periods are common, and where the use of chromates has not been legally restricted yet.

4.5 Recent pretreatment development

4.5.1 Pre-commercial processes [30]

Apart from the transition metal complex technology, other approaches for chromium-free pretreatment have been investigated, including cerate, phospho-molybdate and cobalt amine processes. Most of these are dedicated to Al substrate.

Other investigated processes are based on so-called self-assembling molecules (SAMs). These are linear hydrocarbon molecules with two terminal functional groups that have a bonding capability to both the metal and the subsequent paint coat. Owing to their molecular structure, SAMs assemble to form a monolayer that has a thickness in the order of the hydrocarbon chain length, i.e. 1.5 to 2 nm. These processes currently have found only limited applications, outside the coil coating world, e.g. in the automotive wheel manufacture and in the aircraft industry.

One other class of treatments employs the formation of silane/polysiloxane layers from a sol-gel process. Typical representatives of this class of compounds are terminal amino or glycidyloxy propyl triethoxy silane. The technology requires a hydrated metal oxide surface as obtained by alkaline cleaning. The further process involves hydrolysation of the ethoxysilane functions, bonding to the surface via metal-oxygen-silicon links, and polymerisation/cross-linking of the affixed hydroxy silane moieties to form a polysiloxane. Sol-gel systems have been utilised for a while. They basically form a passive barrier that prevents access of corrosive media, e.g. by hydrophobisation.

Siloxane coatings can also be coupled with the metal surface by plasma or combustion chemical vapour deposition technology. By further treatment with silanes, the functionalised organic groups protrude from the surface, becoming available for bonding with paint coats. For pretreatments in a coil coating process both approaches have not yet taken ground, for reasons of complexity and cost.

4.5.2 *Primer-pretreatment* [31, 32]

The combination of inorganic and organic, polymeric materials can render coatings with additional features apart from adherence and passivation, for instance, fingerprint resistance, lubrication, or priming capability. The idea of combining pretreatment and primer functions in one single stage in the coil coating process has been around for a number of years. For the integrated operation of a coil coating process, such processes allow add-on quality and changes in the manufacturing chain, like the application of extra coats and varnishes.

The primer pretreatment (2-in-1 primer) concept gains a particular edge as it allows the priming process to be integrated at the exit of a coil galvanising line. As well as combining the pretreatment and priming steps, even the cleaning that is normally the mandatory first stage in a conventional coil coating line can be omitted. Even more economic benefit can be generated by an integrated finish coating stage, thereby making storage and transport between galvanising and coating lines obsolete. The effect on the process route is shown in Figure 4.13.

The combined 2-in-1 primer pretreatment concept envisages multiple advancements in coil coating:

- Obsoletion of separate treatment steps in coil coating
- Five-step technology advance
  - Cr-free (heavy metal free)
  - Low-temperature curable
  - Waterborne, enabling solvent-free coating process
  - High yield
  - Option for in-line coating in galvanising lines
- Chemicals available commercially
- Range of suppliers and products growing
- Suitable for compact coil and sheet installations

A primer pretreatment, combined with fast curing techniques, enables process speeds in galvanising and coating to be matched. Solvent-free coating products, moreover, do away with expensive installations for the combustion and thermal destruction of the exhausts from the curing ovens, and drastically reduce the fuel consumption and VOC and CO_2 emissions.

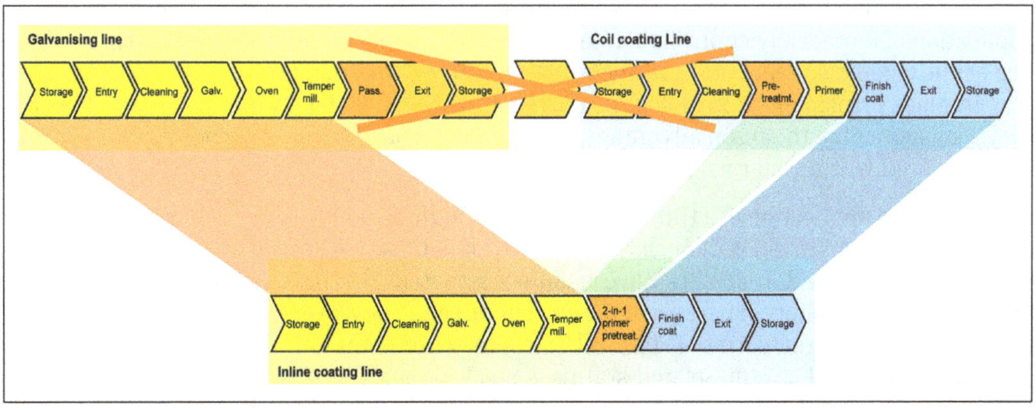

Figure 4.13: Process streamlining by a combined 2-in-1 primer pretreatment concept

The development of a primer-pretreatment therefore should embrace low toxicity by using Cr-free, if possible heavy metal-free chemistry, a low energy consumption by low curing temperatures, a low emission level through a waterbased formulation, high yield by a high solids/reactables content, and the capability for in-line priming in galvanising installations.

There are commercial products available and used in dedicated applications. All relevant suppliers are quoted to have suitable products upcoming, and more universal applications will be possible. The 2-in-1 primer concept is also a perfect match for compact coil and sheet coating installations that comprise start-stop operation and powder topcoating [33].

4.6 Application

4.6.1 Immersion and spray processes

For the application of conventional coil pretreatments [34], immersion or spray techniques are used. Immersion lines have been common for the so-called green chromating (chromous phosphate) pretreatment of aluminium coil.

The drag-out of chemicals is restricted by squeeze rollers at the exit of every active or rinse stage. When properly maintained, the wet film can be minimised to a few millilitres per square metre with these rollers. To avoid damage of the surface of rollers, they have to be lifted off the strip automatically when the welded or stitched joint between two coils passes, and return to their squeeze position immediately after.

Concerning the overflow control and the economy of rinses, the same principles apply as for a cleaner operation (cf. Chapter 3)

4.6.2 Alternative application processes

For certain treatments for steel and galvanised steel coil, the so-called **spray-cell** technique has been in use. Such equipment comprises an enclosed compartment where a high excess of a concentrated pretreatment solution is sprayed onto the coil at the cell entrance, staying there for a reaction time of several seconds. At the cell exit, excess and reaction by-products are removed from the coil surface by a squeeze roller. Poor economy and high pollution brought an end for this technique. Ordinary spray technique has proven to be more convenient and eco-efficient, also saving operation time by bath life control measures.

Over the last twenty-five years, coil coating pretreatments have been adapted to the preferred application method for **no-rinse technology**. As mentioned before, this involves the application of a precisely controlled amount of pretreatment chemicals to the strip surface. The acidic chemicals react with the metal surface, film formation taking place while drying (dry-in-place principle). There are no reaction side-products, and a water rinse is not necessary. Consequently, the main advantages are found in the efficiency of chemical usage and in substantial waste water savings.

The controlled application of a thin liquid film is brought about by a roll-coater. The principle of a roll-coater and its operation, as well as design details are described for the paint coaters in Chapters 6.4. Roll-coaters have been constructed in different ways. When adapted to application of waterborne pretreatment chemicals, they are often also referred to as chemcoaters. For pretreatment, two-roller coaters are common. Roller surfaces usually are made from hard chrome-plated stainless steel, preferably with a grooved surface that allows enough liquid to be transferred. The applicator roll needs to be acid-proof, as well. For illustrations of chem-coaters, see Chapter 1, Figure 1.3, and Chapter 6.1.2, Figure 6.3.

Table 4.6: Safety data and labelling for a chromium-free (left) and a chromate conversion coating (right)

Chromium-free pretreatment	Chromate containing pretreatment
Hazard symbols: C Corrosive	Hazard symbols: T Toxic C Corrosive N Environmentally harmful
Hazardous Substance: Hexafluoro zirconic acid	Hazardous Substance: Cr(3) chromate Cr(6) oxide
R clauses: R 34: Causes burns	R clauses: R 45: May cause cancer R 22: Hazardous when swallowed R 35: Causes severe burns R 43: Sensibilisation possible upon skin contact R 50/53: Very toxic for aquatic organisms, may have long-term harmful effects on waters
S clauses: S 26: Upon eye contact, immediately flush thoroughly with water get medical attention S 36/37/39: During work, wear suitable protective clothing, gloves, eye/face protection S 45: In case of an accident or when feeling sick, immediately see a doctor (if possible, show this label)	S clauses: S 53: Avoid exposition – get special instructions before use S 26: Upon eye contact, immediately flush thoroughly with water get medical attention S 36/37/39: During work, wear suitable protective clothing, gloves, eye/face protection S 45: In case of an accident or when feeling sick, immediately see a doctor (if possible, show this label)
National regulations (Germany): WGK = 2, water polluting product, Classification according to Mixing Rule VwVwS May 17th, 1999	National regulations (Germany): WGK = 3, highly water polluting product, Classification according to Mixing Rule VwVwS May 17th, 1999

Also simpler equipment has been used, involving the controlled spreading of a small amount of the pretreatment liquid that is applied onto the strip a short distance before squeeze rollers. The most recent development involves spraying of the chemicals onto the strip and removing the unreacted excess by low vacuum suction [35].

4.7 Environmental considerations

4.7.1 Legislation

As mentioned before, the European Union (EU) has issued strict legislation in order to control and restrict environmental and health hazards related to the use of chemicals [36]. Some of the related laws and directives have their worldwide counterparts, like the End of Life Vehicles Directive (ELV), the Waste Electrical and Electronic Equipment Directive (WEEE) and the Regulation on Hazardous Substances (RoHS), which have resulted in an effective restriction of the use of chromates and other toxic substances in the respective target markets. On top, the new codex REACH, whose acronym stands for registration, evaluation, authorisation and restriction of chemicals, rules the use of all chemical substances that are manufactured in or imported into the European Union. It is governed by the EC Regulation 1907/2006. All chemicals require official registration and, above threshold levels of use, also authorisation. They are subject to assessments of their exposure risks to health and environment. Chemicals of very high concern, e.g. CMR labelled (carcinogenic, mutagenic and toxic to reproduction), and highly toxic substances like chromate compounds are likely to be effectively banned.

4.7.2 Environment: Labelling and emissions

Also labelling instructions are advantageous for a chromium-free process, owing to its much lower hazard potential. Moreover, when solvent-free coating products can be used, expensive installations for combustion and thermal destruction of exhausts from curing ovens become obsolete, and fuel consumption and VOC and CO_2 emissions are substantially lower. Table 4.6 gives a comparative example of safety data and labels that have to be displayed for a Cr-free vs. a chromate pretreatment process.

4.7.3 Waste Treatment

Waste water that is generated in a chromating operation (treatment baths and rinses) must be processed via a detoxification of chromate ions, e.g. by reduction with sodium bisulphite in acidic solution. After neutralisation with lime and precipitation, the concentration of heavy metals that are present, like chromium (III), zinc, and others, as well as fluoride can be reduced below the required legal limits. The detoxification also builds up a high sulphate level. If limits for direct effluent discarding apply, the sulphate content must be removed by precipitation with calcium aluminate.

Waste water treatment for chromium-free processes is simplified to a treatment with lime. The sludge must be dehydrated and deposited in a landfill site. The cost for treatment and disposal of a typical chromating process is five to eight times as much as with a chromium-free process.

4.8 Literature

[1] Sander, J., Kirmaier, L., Manea, M., Shchukin, D., Skorb, E., Anticorrosive Coatings, Vincentz Network, Hannover 2010, p. 18

[2] Sander, J. et al. [1], pp. 139 f

[3] Sander, J. et al. [1], pp. 137 ff

[4] Lehmann, V., Der moderne Hochofenprozess (Modern Blast-Furnace Process), Naturwissenschaften im Unterricht Chemie, vol. 4, Friedrich-Verlag, Seelze 1993, p. 4

[5] Meuthen, B., Jandel, A.-S., Coil Coating, 2nd ed., Vieweg, Wiesbaden 2008, p. 47 ff

[6] Townsend, H. E. (ed.), Outdoor Atmospheric Corrosion, Am. Soc. f. Testing and Materials (ASTM) Intern., W. Conshohocken PA, USA, 2002, p. 219

[7] Prosek, T., Larché, N., Vlot, M., Goodwin, F., Thierry, D., Corrosion Performance of Zn-Al-Mg Coatings in Open and Confined Zones in Conditions Simulating Automotive Applications, Mat. Corr. 61, 2010, pp. 412 ff

[8] anon., Roheisen- und Stahlerzeugung (Crude Iron and Steel Making), VDEh German Steel Inst., Düsseldorf 2013; www.stahl-online.de/Deutsch/Linke_Navigation/Technik_Forschung/Produktionsverfahren/Roheisen-_und_Stahlerzeugung.php (20-06-13; 11:02 h)

[9] anon., Steel Processing Overview, World Steel Association, Brussels 2012; http://steeluniversity.org/content/html/eng/default.asp?catid=3&pageid=650269858 (20-06-13; 11:11 h)

[10] Sullivan, J., Mehraban, S, Elvins, J., In situ monitoring of the microstructural corrosion mechanisms of zinc-magnesium-aluminium alloys using time lapse microscopy, Corr. Sci. 53, 2011, pp. 3208 ff

[11] anon., MagiZinc, Brochure, Tata Steel, London 2011

[12] Schulz, J., Werner, B., Coil Coating and Zn-Mg coatings, ECCA General Meeting, Proc., European Coil Coating Association, Brussels 2012

[13] Schäfer, J. H., Stoffstrom Aluminium - vom Bauxitabbau bis zum Recycling (Material Flow Aluminium - From Bauxite Mining to Recycling), Gesamtverband der Aluminiumindustrie GDA, Düsseldorf 2008; www.aluminium-ressourceneffizienz.de/media/Stoffstrom_Aluminium.pdf (20-06-13; 11:16 h)

[14] anon., Aluminium, GDA, Düsseldorf 2009, www.aluinfo.de/index-php/alu-lexikon.html (20-06-13; 11:17 h)

[15] anon., About Aluminium - Production Process, Europ. Aluminium Assoc., Brussels 2009; www.alueurope.eu/about-aluminium/production-process/ (20-06-13; 11:19 h)

[16] anon., About - Metal Production, Novelis Inc., Atlanta GA, USA 2013; http://www.novelis.com/en-us/Pages/Metal-Production.aspx (20-06-13; 11:36 h)

[17] anon., Rolling Aluminum: From the Mine Through the Mill, 3rd ed., Aluminum Association, Arlington, VA, USA, 2008

[18] Sander, J. et al. [1], p. 140

[19] Sander, J. et al. [1], p. 141

[20] Sander, J. et al. [1], p. 18

[21] Sander, J. et al. [1], p. 142

[22] Sander, J. et al. [1], pp. 143 ff

[23] Androsch, F. M., Stellnberger, K.-H., Wolpers, M., Jandel, L., Drescher, D., Sander, J., Seidel, R., Chromate-free Coil Coating and One Year of Production Experience, ECCA General Meeting, Proc., European Coil Coating Association, Brussels 1999

[24] Schinzel, M., Chrome-free Systems for Coil Coating, ECCA Autumn Congress, Proc., European Coil Coating Association, Brussels 2006

[25] Sander, J., Chromatfreie Systeme für Coil Coating - Stand der Erfahrungen (Chromate-Free Systems for Coil Coating - Status ...), Europ. Coil Coating Assoc. - German Group, Workgroup Proc., Solingen 2002, cit. re. Schinzel [24]

[26] Sander, J. et al. [1], pp. 113 ff

[27] Androsch et al. [23], Chapter 3

[28] Klimow, G., Fink, N., Grundmeier, G., Electrochemical Studies of the Inhibition of Cathodic Delamination of Organically Coated Galvanised Steel by Thin Conversion Films, Electrochim. Acta 53, 2007, pp. 1290 ff, 1291

[29] Lienkamp, P. (ed.), Sander, J., Körner, T., Kleinschmidt, T., More than 10 years of experience with chrome-free pre-treatment of aluminium - Perfect protection against corrosion, Press release, Henkel AG & Co.KGaA, Düsseldorf 2008, cit. Int. Aluminium J. 84 (11), 2008, p. 36 ff

[30] Sander, J. et al. [1], p. 146

[31] Sander, J. et al. [1], p. 147

[32] Sander, J., Novel Surface Treatment of Metal Strip in the Coil Coating Process, Millennium Steel 2009, pp. 146 ff, 148 ff

[33] Bielefeld, F. W., Sander, J., The profitable choice: From small-batch post-painting to Compact Coil Coating, European Coatings Market Day Coil and Can Coatings, Proc., Vincentz, Hannover 2011

[34] Sander, J. et al. [1], pp. 149 ff

[35] anon., Products: Vacumat: Vacumat Special Coat, Schiele Maschinenbau GmbH, Niederzissen 2011; www.schiele-maschinenbau.de/... (20-06-13; 12:14 h)

[36] Sander, J. et al. [1], pp. 147 ff

5 Coil coating paints

5.1 Definition

Organic coatings can be defined as liquids, pastes or powders that enable transparent or opaque coverings providing decorative, protective and specific technical properties. They comprise organic film formers, a carrier medium (in case of liquid coatings), pigments, fillers and a range of additives [1].

Their major features include

- the **ease of handling** such as a delivery form that is ready to use, stable and quickly redispersible upon agitation
- **good applicability** in terms of a suitable rheology, wetting behaviour and film formation
- **good final properties**, as colour retention, curability, mechanical and chemical properties incl. weatherability, and, last but not least
- little impact on the **environment**

5.2 Ingredients: Carrier media

In liquid solution coatings, the role of the carrier is to solvate, i.e. disentangle and dissolve polymer molecules individually, as well as to control the viscosity of the liquid.

Most paints in coil coating are solventborne, and a variety of different **organic solvents** are used, mostly blended to match the solubility required by the polymers. Solvents may be either polar (alcohols, ketones, glycol ethers, esters), or non-polar (aliphatic, aromatic hydrocarbons).

Water is used as the carrier in dispersion formulations. As it is polar by nature, it requires the polymeric resins to be rendered hydrophilic by suitable functionalisation. Sometimes, also co-solvents are added (usually high-boiling organics that are not regarded as volatiles).

Table 5.1 displays physical features and typical applications of carrier media used in coil coating paints [2–14].

5.3 Drying and curing mechanisms

5.3.1 Physical drying

Organic coatings are also categorised regarding their drying and curing mechanisms.

During physical drying of solvent-based paints, the evaporation of the carrier leads to closing up and mechanically entwining of the individual polymer strands of the binder mole-

Table 5.1: Physical data and typical uses of selected carrier media for coil coating paints[3-14]

	Water	n-**Butanol** butane-1-ol	**Butyl glycol** 2-butoxy ethanol	**m-Xylene** 1,3-dimethyl benzene	**Petrol 100/140**	**Solvesso 150**
Boiling point/range at 1 bar [°C]	100.0	117.7	168.4	139.1	100 to 140	181 to 207
Evaporation enthalpy at bp [kJ/kg]	2259	615	368	391	336	361
Rel. evaporation rate (diethyl ether = 1)	80	33	163	13.5	~7	~43
Density d_{20} [g/cm³]	1.0000	0.8098	0.9015	0.8684*	0.695 to 0.726	0.838 to 0.905
Surface tension at 25 °C [mN/m]	71.97	24.93	26.14	28.47	~18	~28
Carbon content [%]	0.0	64.8	61.0	90.5	84.1	89.3
Refractive index n_D^{20}	1.3299	1.3993	1.4198	1.4978	~1.4	~1.5
Typical use in coil coating	water-borne paints	amino-phenolic resins	polar resins, co-solvents for aqueous paints	particularly suited for polyesters	unpolar resins, e.g. polyvinyl ethers	particularly suited for polyesters, acrylics

** 15 °C*

cules. Thereby they form a new spatial structure (conformation) that is supported by hydrogen (bridge) bonds. Thus a closed film is generated that usually has good cohesion features and shows thermoplastic behaviour, i.e. softening/liquefying with increasing temperature.

With a few exceptions, physically dried coatings have only limited chemical fastness, in particular no resistance to solvents.

Physical drying of aqueous dispersions requires, after the evaporation of water, the individual, spherical particles (micelles) of the dispersion to close up and get densely packed. This is only possible when the dispersion is still suitably fluid, and therefore depends on the temperature. A minimum film forming temperature (MFFT) is characteristic of each formulation. The film adopts a structure which has the dispersion particles hexagonally stacked, with ideally no interstices between them. Cohesion is interpreted in terms of the reduction of total surface, and the gain in surface energy aligned with it. When present, inorganic compounds assemble along the interfaces of compressed particles. Reactive inorganics, e.g. chromates, may form linking bonds between particles, and also hydrogen bonding increases the cohesive force.

At temperatures above MFFT, the dispersion micelles finally fuse into a uniform film that provides enhanced cohesion and mechanical fastness. Emulsifiers which are usually present to control the size and distribution of the micelles in the dispersion are either alloyed into the film or expelled from it. The first option leads to a softening effect. The latter may have effects on e.g. the adhesion or gloss of subsequent coatings, when the emulsifier is not evaporated [15-20].

5.3.2 Chemical curing, crosslinking

The formation of an infinite three-dimensional polymer network is necessary to obtain chemical and higher mechanical resistance. Through the reaction of suitable end or side-chain groups of the polymers additional crosslinking bonds are formed. The polymer size and molecular mass thus grow by orders of magnitude. This enlargement of polymers involves addition, or condensation mechanisms.

Linear growth of the polymers renders a thermoplastic material that is sensitive to temperature increase, as it softens and finally liquefies. On the contrary, the three-dimensional crosslinking yields thermosetting films or duroplastics. A characteristic property of thermosetting films is the so-called glass transition, where the polymer film changes from glassy to rubbery behaviour. This change of state is aligned with a specific glass transition temperature, T_g. As the resin is usually composed of a blend or a range of polymers with a distribution of molecular weight, in practice, the T_g is not a discrete temperature, but a narrow temperature interval. By convention, the T_g onset or midpoint temperatures can be given. These elasticity regimes display typical stress/strain features. The glassy state behaves elastically, however with only a little elongation before rupture occurs. Around the T_g and slightly above, the system usually shows a visco-elastic behaviour, i.e. it is deformed under stress, however returns to its conformation when relaxed. Further above the T_g, deformation is permanent, however the film may rearrange in a different conformation, before, at higher stresses, elongation and finally cleavage occurs [21, 22].

Crosslinking paints cure in a temperature/time window. Typical temperatures and times for the fast coil coating process are 240 °C and some 30 s, to allow for complete evaporation, activation of capped reactive groups and the crosslinking reaction itself to take place [23, 24].

Under-curing leads to soft, non-resistant coatings with poor durability, while over-curing usually results in brittle films with poor adhesion or even thermal degradation indicated by discoloration (yellowing). Cracked and degraded films obviously are also not durable. Hence, proper curing is achieved only at an optimum temperature/time level [25].

5.3.3 Radiation curing systems

Though it has not taken a considerable share of the coil coating paint market, the concept of radiation curing shall be discussed briefly. It might take a more important role in the future, due to its process speed and environmental benefits. It has been argued that a UV or an EB unit are likely to consume much less energy than needed by a conventional convection oven. This is particularly due to the energy that is lost through heat dissipation in thermal curing.

Radiation curing is brought on by ultraviolet (UV) radiation (cf. Figure 5.1), or accelerated electrons (electron beam, EB) from a high-voltage cathode. EB curing is a quite common technology used in laminating and printing of packaging foils, e.g. aluminium-polypropylene compound laminates.

UV curing requires photoinitiators to be present that absorb the radiation and then decompose, via an excited state, into reactive radicals. These attack suitable monomeric unsaturated organic molecules, mostly acrylate derivatives, to start a photopolymerisation reaction. UV is used preferably with transparent coatings, as certain pigmentation obscures the radiation.

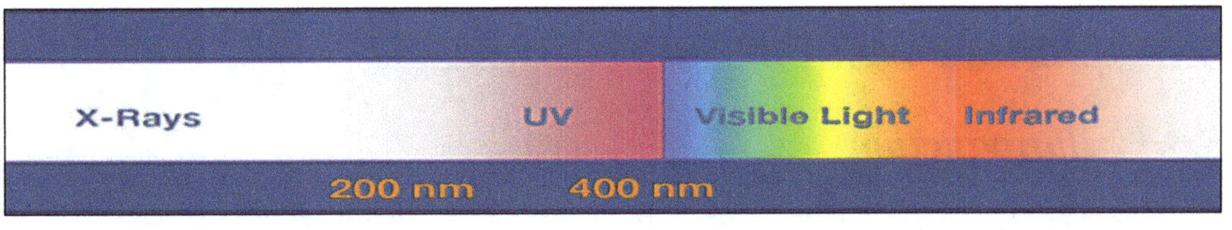

Figure 5.1: Electromagnetic spectrum Source: Henkel

EB can be applied regardless of the pigmentation of a coating, and in any thickness common in coil coating. Hence, the technology appears particularly suited for topcoats. It is capable of attacking C=C double bonds, triggering the polymerisation directly. The residual free radical concentration is very low in EB curing.

The polymerisation in either radiation curing technology is ultra-fast and completes within less than a second, forming cured, crosslinked films in situ. By choice of a suitable formulation, coatings with good adhesion and corrosion protection properties, but also with additional features, e.g. anti-fingerprint, or scratch and mar resistance can be designed. The formulations can be solvent or waterborne, but these carriers must be evaporated thoroughly prior to curing. Hence, most effectively, 100 %-systems are used, that react off entirely [26–28].

Outdoor exposure and 1,000 h QUV-A results of several EB cured urethane and isocyanate acrylate based paints were reported. While the main degradation path in natural weathering was photooxidation of the acrylic backbone, particularly with aliphatic moieties prone to α-H abstraction, urethane and ester crosslinks were intact. These results suggested that further optimisation was worthwhile to pursue [29]. UV curable clearcoats are used on stainless steel coil to obtain a virtually invisible, fingerprint and scratch-resistant coating used e.g. in the manufacture of domestic appliance [30, 31].

5.4 Powder paints

Powder paints are considered to be particularly environmentally friendly, as they allow the coating process to be performed without solvents. They are applied onto cleaned and pretreated metal surfaces, by electrostatic methods (e.g. gun spraying), or as slurries in water.

Powder primers and so called hybrid topcoats are usually based on epoxies and blends of epoxy/acidic polyester, respectively. Admixing of polyester resins improves the resistance against chalking and yellowing when exposed to elevated temperatures. For increased weathering resistance, polyester or polyurethane powders are used. Powder paints therefore contain crosslinking agents as required for the particular binder chemistry, e.g. glycidyl esters, or blocked isocyanates or uretdiones. Thermoplastic powders (polyamide, PVC or PVdF based) are only of minor importance [32].

Coil and sheet for architecture and appliance are sometimes powder coated in specialist lines, usually from conventionally pre-primed coil stock. As the process requires systems that are applied and cured fast, powder applications are yet limited. Fluidising and coalescing (up to 150 °C), and crosslinking of the powder paint (up to 200 °C) usually takes about 30 min. Therefore, new equipment is being developed that allows applying powder in bulk amounts and speeds beyond the conventional electrostatic guns, e.g. electro-magnetic brush, or fluidised bed designs (see Chapter 6.3.2) [33, 34]. Also, particularly efficient curing equipment is necessary to achieve the required speed in film forming and crosslinking. Approaches with infrared and induction ovens, as well as near-infrared (NIR) equipment and preheating devices have been reported [35–37]. Only by these special measures, and with reactive paints, the technology may be enabled for use in compact coil and sheet lines [38].

5.5 Binder resins

5.5.1 Introduction

The polymeric portion of a paint is usually also called a binder. Pigments and fillers of the formulation are dispersed in the binder resin, which means that the individual solid particles are entirely surrounded by resin molecules that keep them apart from each other, thereby preventing them from agglomerating and flocculating. Surfactants (dispersants) are used to help in separating and wetting the solid particles with the resin. When the binder becomes solid itself by the action of physical drying or chemical curing, it entraps the pigment and filler particles. The chemical and physical properties of the resin rule the overall features of the properly formed film, be it mechanical, adhesive and cohesive characteristics, the chemical or the weather resistance of the coating.

There are a variety of binder chemistries available. The most important for coil coating are polyesters, polyurethanes, PVdF or PVC for topcoats, or epoxies for primers and backing coats. Regarding their film forming mechanism, the binders can be grouped under the thermoplastic or thermosetting classes of binders [39].

5.5.2 *Polyvinyl chloride, PVC*

Polyvinyl chloride is commercially obtained by suspension polymerisation starting from vinyl chloride. Most material is further processed by radical chlorination to yield a polymer containing up to 65 % chlorine. The typical molecular weight of commercial polyvinyl chloride is around 70,000 u. Such material has a very poor solubility. Improvements in this respect can be achieved by copolymerisation with small amounts of other monomers such as vinyl acetate.

In coil coating, PVC is applied as a plastisol, i.e. a dispersion of PVC powder in suitable plasticisers. During heating, the resin dissolves in the plasticiser, rendering a solid, thermoplastic material upon cooling. Pigments are introduced by mixing with the plasticiser, before the blending with the resin takes place. The pigment concentration is often lower than in polyester paints however this is compensated for by the higher film thickness.

Frequently, mixtures of finer and coarser PVC resin preparations are employed. The finer suspension grade resin is used as the main component, while smaller portions of a coarser emulsion resin are added to reduce the gloss of the resulting coating [40, 41].

PVC is subject to multiple thermal or radiation-induced decay reactions. Heat will cause rapid dehydrochlorination that is even autocatalysed by the released hydrochloric acid. From this elimination reaction, that can already take place during the physical drying (gelation) process, C=C double bonds result in the hydrocarbon backbone. The first visible effect of heat degradation of PVC therefore is darkening (yellowing) of the film, as, with increasing number of C=C double bonds per molecule, conjugated double bond systems will evolve which absorb electromagnetic radiation. Visible light is absorbed by systems of eight conjugated double bonds or more, leading to gradually increasing blackening the more of these chromophores are formed [42]. Additives are used to enhance thermal stability. Instead of the formerly used lead (Pb) and cadmium (Cd) compounds, the much less hazardous barium zinc soaps, in combination with phosphates, were identified as effective materials to inhibit dehydrochlorination. In effect, the chloride is immobilised as low-soluble barium chloride, while the soap buffers the generated H^+ ions. Even more recently, the barium compound is being replaced by calcium zinc soaps, avoiding potential problems with the barium labelled as a heavy metal [43].

UV irradiation also causes destruction of the PVC polymers. UV first induces a homolytic cleavage of the C-Cl bond. The resulting free radical is subject to onward reactions, e.g. oxidation by atmospheric oxygen, or radical recombination (with or without prior H abstraction). The degradation products generally exhibit a higher brittleness, hence lesser mechanical and weathering performance. To block this decay path, UV absorbing additives are used.

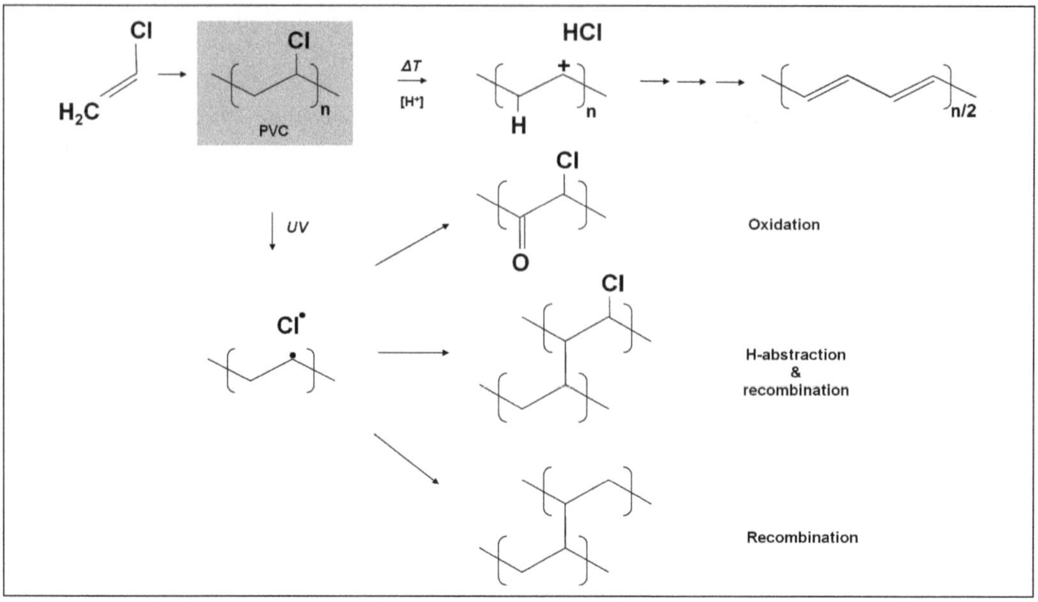

Figure 5.2: Generation of PVC and major decay paths. Top: Thermal, acid-catalysed; bottom: UV induced

Figure 5.2 shows the reaction sequence for the generation of PVC from the vinyl chloride monomer and several decay paths that are induced thermally or by UV radiation.

For many years, phthalic acid esters, like dioctyl phthalate (DOP) or diisodecyl phthalate (DIDP), have been used as plasticisers in PVC plastisol formulations. Recently, they have come under scrutiny due to possible health hazards including carcinogenity and endocrine disruption (fetal defects). As a suitable replacement, aliphatic materials such as adipates have been used as plasticisers [44]. Only a very recent change in the US chemical policy might attenuate the critical view on phthalates [45].

A PVC plastisol coating typically has a gauge of up to 200 μm. The coatings are used for heavy duty applications that require good water, solvent and fuel resistance, and also good anticorrosion properties. The thermoplasticity allows the creation of surface textures by embossing of the soft coating while it is still warm.

5.5.3 Polyvinylidene difluoride, PVdF

Polyvinylidene difluoride, PVdF, is mainly produced by catalytic polymerisation of 1,1-difluoro ethene, from precursors trichloro ethane and hydrofluoric acid (HF), as illustrated in Figure 5.3. It is a thermoplastic, solid polymer that is UV stable by nature, and therefore does not chalk nor dull under weather exposure [46]. Its limited mechanical strength makes it but suited for wall cladding purposes. Its main uses are on architectural aluminium.

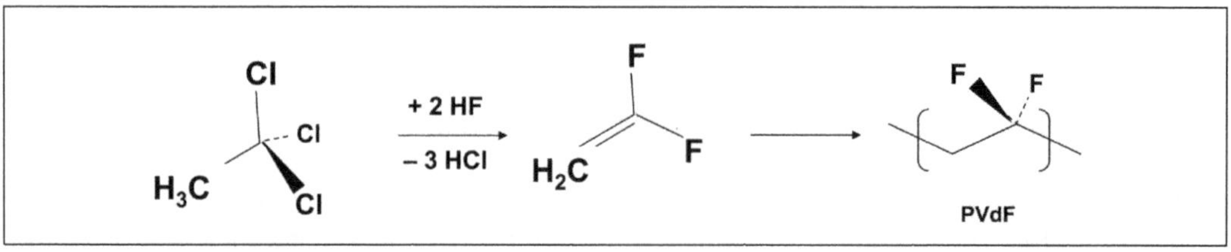

Figure 5.3: Generation of polyvinylidene difluoride (PVdF)

In a typical formulation, PVdF and acrylic resins are blended to enable colour pigmentation, the pigments being dispersed in the acrylic part, and to shift the T_g to a suitable range. The standard peak metal temperature (PMT) of PVdF coatings is around 249 °C, allowing for the PVdF and acrylic resins to efficiently form an alloy that results in a tough, chemically resistant film. On cooling, the in-situ coating has a T_g of 20 to 30 °C, inbetween the low T_g of the PVdF and the high T_g of the acrylic resin. The PVdF molecules also align during the cooling, to form crystallites which have a fusion point around 158 °C. This semi-crystalline structure is likely to be one reason for the toughness of the coating and, to a certain extent, the low gloss levels achievable.

The strong C-F bond has no absorption in the UV-A or UV-B parts of the spectrum. Hence it is highly durable. The acrylic resin, on the other hand, is less stable and so the molecules closest to the surface will suffer from photooxidation. However, the remaining PVdF moieties tend to reflow into the gaps, essentially sealing the system, resulting in the extended lifetimes associated with PVdF [47].

In the past, primers for PVdF systems were typically prepared from epoxy resins, while more recently, polyester resins have taken ground. When used for non-food applications, chromate pigments were in common use in these primers until the threat of legislation along with advances in chromate free technology forced the hands of the paint producers.

5.5.4 *Epoxy resins*

Epoxy resins are generated from OH functional backbones that are etherified with epichlorohydrine. Essentially, during this etherification, the glycidyl ring of the epichlorohydrine is opened in the reaction with the OH group, thereby forming a secondary OH group adjacent to the chlorine. When NaOH is present, HCl is eliminated, and a glycidyl group reformed [48]. Typical precursors are aliphatic, cycloaliphatic, or aromatic molecules, e.g. bisphenol A (BPA). The latter gives rise to bisphenol A diglycidyl ether (BADGE) and its higher homologues, as depicted in Figure 5.4. As can be seen from the diagram, these higher homologues have secondary OH groups in their backbone that is formed by n-fold repetition of the basic sub-molecular moiety. In a common range of commercial epoxy resins, the average value of n is 12. Hence, such material also has 12 secondary OH groups per molecule, giving rise to a resin that reacts predominantly through them and not through the epoxy or glycidyl group. BPA has been the most important source, however now is sought to be replaced, being suspect of having negative effects to health and environment, e.g. mutagenic or endo-

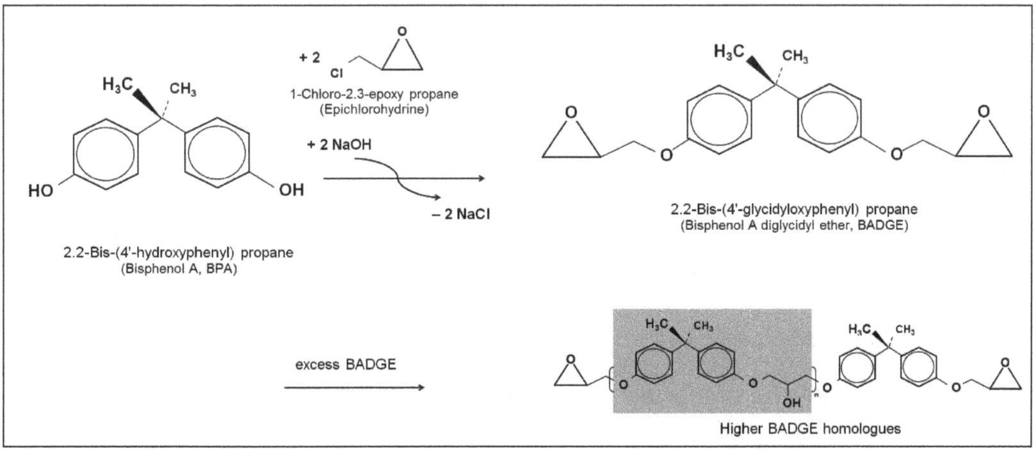

Figure 5.4: Generation and oligomerisation of bisphenol A diglycidyl ether (BADGE)

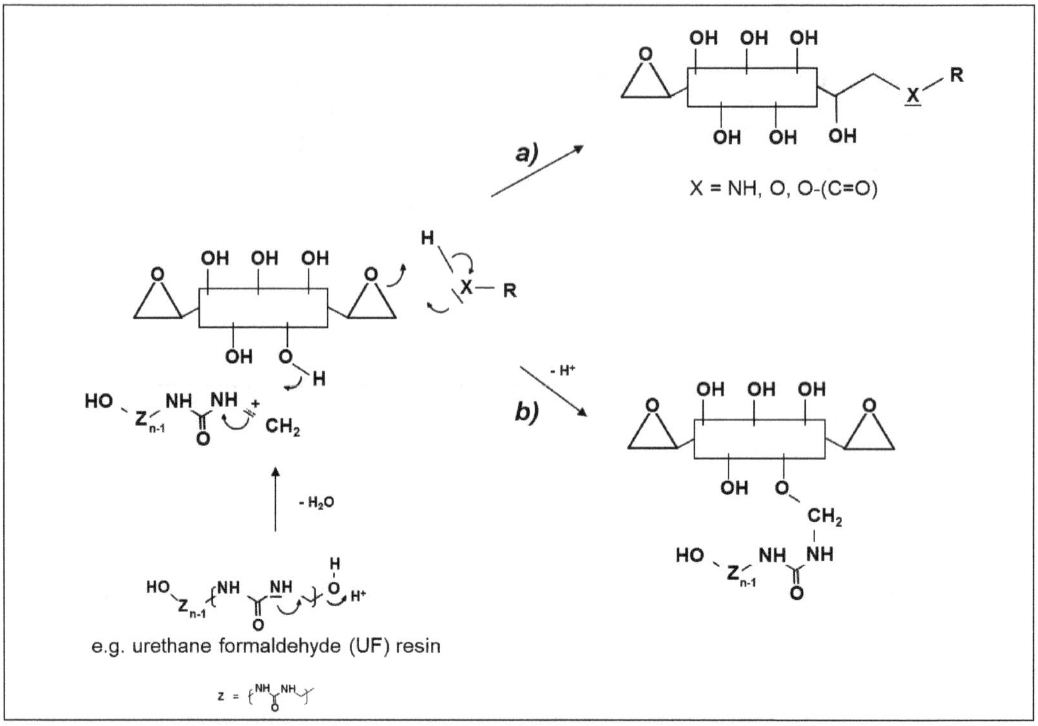

Figure 5.5: Principle of crosslinking reactions with epoxy resins: a) Ring-opening addition at the oxirane group; b) Etherification of a secondary hydroxyl group by a nucleophilic substitution reaction

crine defects [49]. It is reported only just in these days that the US Environmental Protection Agency might refrain from having BPA listed as a chemical of particular concern [50].

The formation of the polymer is achieved by opening of the epoxy or oxirane ring and subsequent polyaddition, or by reaction with diols (in the so-called advancement process) [51]. Preferred crosslinkers for epoxy resins that link via the oxirane ring are amines, carboxylic acids, and anhydrides, like phthalic anhydride. Urethane formaldehyde resins have been used particularly in coil and can coatings, as they efficiently crosslink via the secondary OH groups [52, 53]. Exemplary reactions that are used for modifications and crosslinking of epoxy resins are shown in Figure 5.5.

Epoxy coatings provide high barrier properties, good adhesion and mechanical fastness. Their flexibility is very much dependent on their degree of crosslinking. Epoxies, in particular aromatic epoxies, are prone to light-induced degradation, hence show strong chalking, discoloration, and loss of gloss upon UV exposure.

Reactions with carboxylic acids, in particular native fatty acids, yield a range of epoxy esters that enable different rheology features and performance, in particular improved flexibility. Due to their poor UV resistance but high barrier and adherence properties epoxy paints are used as primers and backers in coil coating, or as lacquers and varnishes in can coating.

5.5.5 Polyester resins

5.5.5.1 Raw materials for polyester resins

Polyesters are a very versatile class of resins for both the coil and can industries. They dominate the volumes of paint sales to coil coaters (cf. Chapter 5.6 below). The resin manufacturers can adjust the backbone architecture, molecular weight and functionality, so that polyester resins can be tailored for any end use.

Polyesters are formed from a polycondensation reaction between diacids, and di or triols. When water dispersability is required, also triacids such as trimellitic acid are used. In order to drive the condensation to completion, the water that is liberated during the course of the reaction is distilled off the reaction vessel, pulling the equilibrium to the desired end products.

Conveniently, anhydrides are employed instead of the acids, thereby reducing the amount of water that is present in the reaction mix beforehand. Basically, also esters or cyclic esters (lactones) can be used as starter materials. The former require the liberated alcohol, e.g. methanol, to be removed from the reaction pot, the latter give access to aliphatic polymer blocks only. The principle of the polycondensation reaction is illustrated in Figure 5.6.

The most common diacids are the aromatic, phthalic acid isomers (ortho-, iso-, terephthalic acid) or (aliphatic) adipic acid. The glass transition temperature, T_g, and thus the ultimate properties of the film at its operating temperature are controlled by balancing the proportion of aromatic to aliphatic acids. A greater proportion of aromatic acids leads to a higher T_g. The choice of aromatic acids is made according to the desired properties and the price-performance balance. Orthophthalic anhydride is the most common source due to its lower cost. Isophthalates provide the highest durability due to their low UV absorbance, while terephthalates are the most formable grace to their quasi-linear structure.

The diols are often selected from ethylene glycol, 1,2-propanediol, 2,2-dimethyl propane-1,3-diol (neopentyl glycol, NPG), and hexanediol. Again the question of which is more important, price or properties, largely determines which raw materials are used. Triols are used to introduce branching, and thus increase the number of hydroxyl groups per molecule, which can enhance the polymer's resilience. The most common triol used is trimethylol propane.

The final properties of a paint film are determined by the backbone architecture of the resin, the degree of crosslinking and the molecular weight. This final parameter is adjusted by formulating to the correct stoichiometry, and by controlling the extent of the reaction. High molecular weight polyesters (~20,000 u) have been used in primers and highly formable domestic appliance coatings in the past. The drawback is the high viscosity these resins display when dissolved in organic solvents, hence limiting the operable solid content. Mid-range molecular weight polyesters are most commonly used for topcoats, as they are seen to give optimum properties. Low molecular weight resins can be employed to increase the solids content, hence also giving better coverage, however they tend to result in soft films with poor mechanical performance [54, 55].

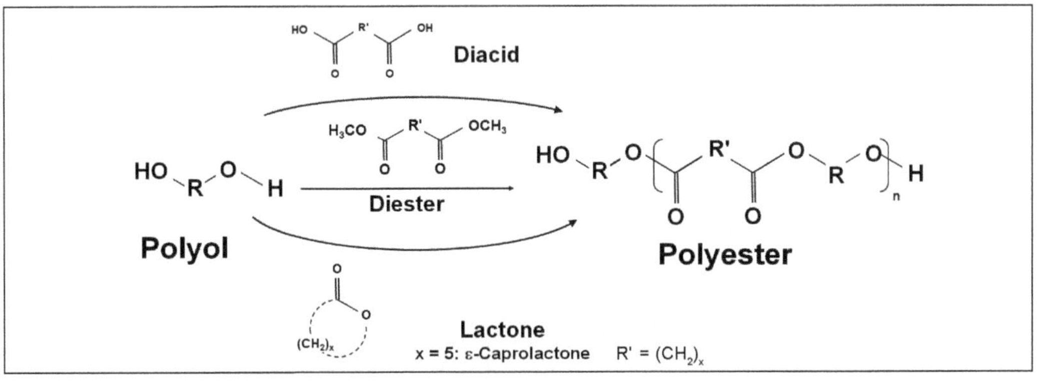

Figure 5.6: Access routes to polyester resins by reaction of polyols with a) diacids; b) diesters; c) cyclic esters (lactones)

5.5.5.2 *Crosslinking of polyesters with melamines*

On their own, polyester resins vary from being slightly sticky (low molecular weight) to being hard (high molecular weight) in their natural undissolved state. They need to be crosslinked if they are to work as protective coatings. As crosslinkers for polyesters in both coil and can coatings, most commonly multifunctional melamine derivatives are chosen, mainly due to their cost. Trimellitic anhydride is used in reaction systems yielding high-solids polyesters [56].

The most commonly used melamine derivative is hexamethoxy methyl melamine (HMMM). Figure 5.7 represents a typical crosslink in a polyester polymer network, showing the HMMM in an idealised form, as the degree of methoxylation of this melamine-formaldehyde resin is seldom exhaustive. Some of the end groups terminate as hydroxymethyl ($-CH_2-OH$) or simple imino (=NH) groups, and occasionally, two molecules are joined together through a nitrogen bridge. There are several variations of commercial HMMM materials, in which the proportions of the OH, NH functional, and dimerised molecules change depending on how fast a reaction is required. The often assumed hexafunctionality would mean that anything between 3 and 7 % of the melamine resin was sufficient to match the hydroxy functionality of the polyester resin. However it has been discovered empirically, that melamine concentrations of between 10 and 15 % are required for rapid reaction. The highly methoxy functionalised melamines are more durable than the resins with a larger number of hydroxy or imino groups, partly because the latter also have a tendency to facilitate moisture-enhanced photooxidation.

It has been shown that melamine molecules tend to spontaneously assemble, within a time-frame of picoseconds (10^{-12} s), in small agglomerates around the OH groups of the polyester resins. These agglomerates are around 5 nm in diameter thereby setting up a nanoparticulate system, which might explain why polyester melamine coatings have the overall good performance that is observed [57].

During crosslinking, the T_g rises as the crosslink density increases, though this effect tails off as full cure is approached. This is when the 3-D network has been widely established and virtually all of the available methoxy groups have reacted.

One popular way of establishing whether or not complete cure has been achieved is the rub test with methyl ethyl ketone (MEK, cf. Chapter 7.2.2.2). Usually, an arbitrary number of 100 MEK double rubs is taken to represent full cure. However, there is some discussion about whether or not this number is correctly chosen. It is argued that a real exhaustive cure is not obtained at this stage, and MEK double rubs could be increased up to 600 and beyond. Obviously, the effect of increasing the cure is to enhance hardness and reduce flexibility. It is also shown that the durability is much better when greater degrees of curing are achieved during the reaction, however the authors also demonstrated that the cure can go too far, and beyond a certain optimum, the properties start to fall away. It shall be pointed out that these investigations do not refer to destructive thermal overcuring by exaggerated temperature [58].

The reaction between HMMM and the hydroxyl group is an acid catalysed transetherification. This involves a nucleophilic substitution reaction rendering the larger ether with the melamine moiety affixed and methanol, as a small leaving molecule. Methanol usually is evaporated under the curing conditions, however it might also be entrapped in the coating, thereby causing boil defects. This phenomenon generally restricts polyester-melamine coatings to dry film thicknesses of around 20 micrometres. The remaining methoxy groups of the melamine can be used to react with hydroxy groups on other polyester chain ends, and thus produce a crosslink as illustrated in Figure 5.7.

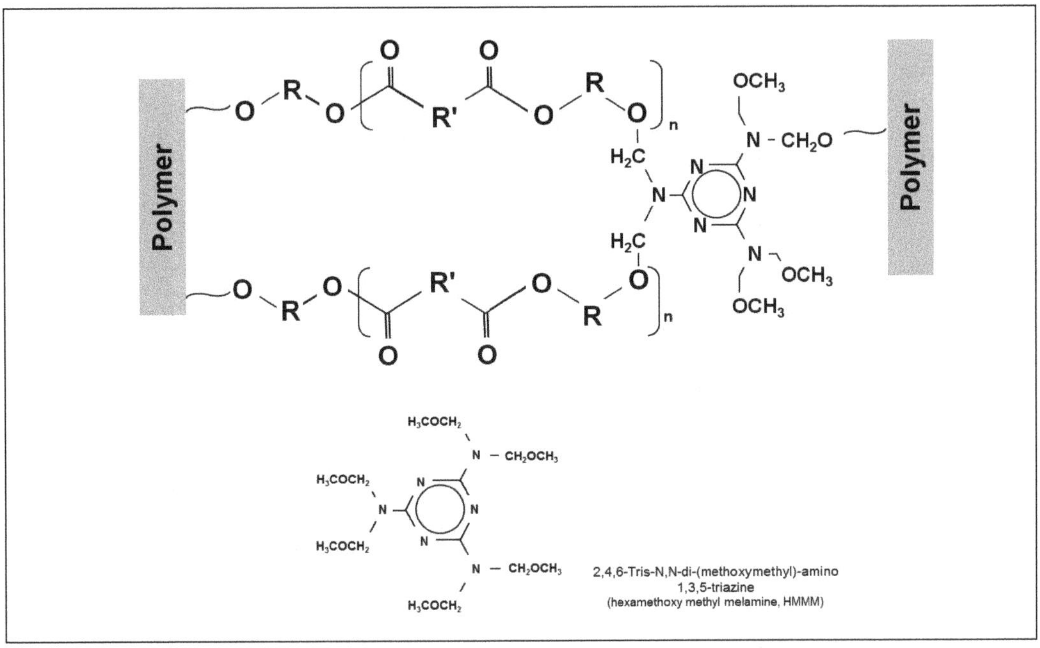

Figure 5.7: Typical crosslink of two polyester strands by condensation with hexamethoxy methyl melamine. The remaining methoxy groups can be used for similar reactions, as long as they are accessible for hydroxy groups of the polymer

The acid catalyst used in the process can be obtained in a plethora of different formats. The most common are sulphonic acids of toluene and naphthalene. These materials can be prereacted with amines to produce a blocked catalyst that only becomes active above temperatures of about 120 to 140 °C. In this way stable formulations can be manufactured and delivered to the customer. Judicious choice of catalyst will however still result in sufficiently rapid cure on line. The catalysts are often supplied in alcohols which dissolve the acid, but are also compatible with the other components of the polyester formulation.

The curing process can be very rapid indeed. NIR and induction curing can raise the temperature of the paint and substrate from room temperature to e.g. 230 °C in less than 3 s. This does not mean, in general, that full cure is achieved in this time frame, because the kinetics demand the reaction to take a certain time even at these temperatures. Nonetheless, dwell times of little more than 7 s have proven possible. Most modern coil lines still are operated using dwell times of 20 to 30 s, with peak metal temperatures of around 224 to 232 °C. It is only aluminium lines that can achieve shorter dwell times with conventional convection ovens. Hence, NIR and induction equipment offers an opportunity to arrive at shorter processes and line lengths (cf. Chapters 6.4.1.2 to 6.4.1.4). A drawback associated with such fast curing is that reactions may occur too quickly with the risk of methanol being entrapped and the resulting boiling blister mentioned above.

The durability of polyester melamines depends largely on the backbone architecture of the polyester resin. This is because that architecture controls the T_g that should be as high as possible, in order to help cage and immobilise radicals that may form during photooxidative decay processes, and to better resist moisture ingress. Furthermore, the backbone architecture determines the UV/visible absorption characteristics of the polyester melamine system, which has a significant impact on how fast the coating will degrade under the conditions met in the field, i.e. whether UV radiation is intense or slight, and the humidity levels are high or low.

5.5.5.3 Crosslinking of polyesters with polyisocyanates

The reaction of polyester resins with isocyanate-based crosslinkers through their hydroxyl functions yields polyester-polyurethane coatings with their typical features of high durability and temperature insensitivity. These are often referred to as high-durable polyesters (HDPE). It shall be pointed out that ECCA statistics deals with this kind of coating systems as polyurethanes.

The formation of these resins via polyester precursors proceeds along the common routes of polyurethane chemistry. It is therefore described in Chapter 5.5.6.

The polyesters used for this type of coating system are very similar if not the same as those used with polyester-melamine systems, although those with low hydroxyl functionality may not cure too well. Another factor that could impact on the choice of polyester is the use of isocyanates in high-build systems. This is because they have no volatile reaction products such as alcohols, and there is no risk of boiling failures when used at high film thicknesses. Such high-build primers and topcoats are often applied at dry film thicknesses of 40 to 60 µm. Their target application in the industry is the same as for plastisols. By replacing these, the associated use of possibly critical chemicals as PVC and phthalate plasticisers can be avoided.

5.5.6 Polyurethanes

Polyurethane links are formed by reaction of isocyanate and hydroxyl (-OH) functions by way of an addition. The polyols can be of any suitable class of resins like polyethers, polyesters or polyacrylics (Figure 5.8) [59].

As isocyanates are very reactive species that undergo conversion with many different partners (amino, thiol, carboxylate, ether) and are also water-sensitive, they are usually blocked to obtain stabilisation. Reaction with water results in the formation of an N-substituted carbamic acid that decomposes, liberating the lower amine and carbon dioxide.

In order to ensure the crosslinking does not occur until the coating is in place, the isocyanate is blocked with compounds such as methyl ethyl ketoxime, caprolactam or dimethyl

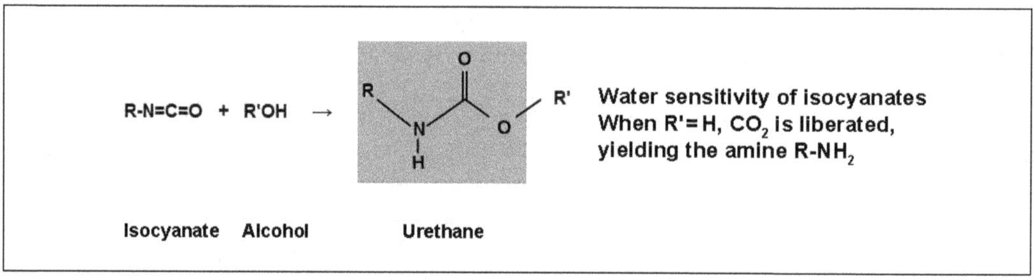

Figure 5.8: Formation of the urethane bond between isocyanate and alcohol

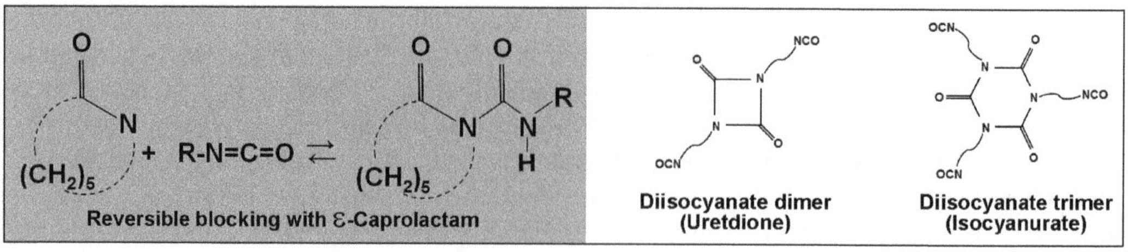

Figure 5.9: Reversible blocking of isocyanates

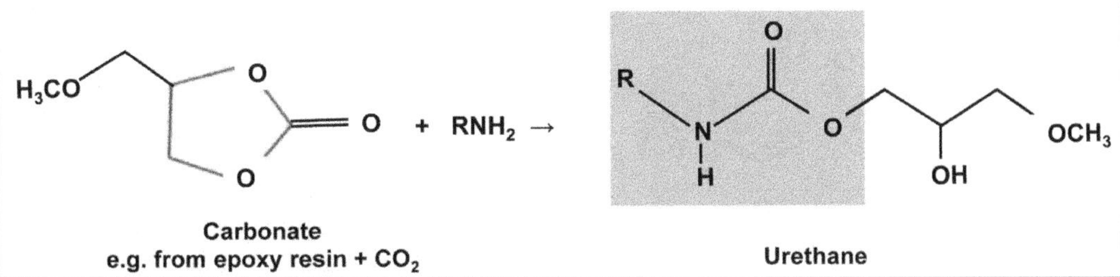

Figure 5.10: Formation of a polyurethane without toxic isocyanate

pyrazole. These molecules unblock from the isocyanate at different temperatures, i.e. 140 to 160 °C, 160 to 180 °C, and 110 to 120 °C, respectively, giving the formulator some latitude in the desired sure response.

Once activated, the isocyanates react quickly, however overall on standard lines, the dwell times involved in polyester-isocyanate systems tend to be longer than those for polyester-melamines.

Regarding the supply of isocyanate crosslinkers, the formulator has a plethora of commercial materials to choose from. As per normal, this choice depends on the properties required of the final coating in service. But also safety and health concerns play a role, of course. Simple monomeric isocyanates can cause respiratory problems, and hence are not used in coil coatings. Instead, the isocyanates are provided as dimers (uretdiones) or, more commonly, trimers (isocyanurates, see Figure 5.9). This oligomerisation moreover helps to block the free isocyanate and avoids unwanted and premature reactions [60].

These multifunctional blocked isocyanates are incorporated into the liquid formulation in the normal way. Usually, they have been reacted, assisted by tin catalysts, on heating. The use of the tin compounds, however, is being phased out due to their adverse environmental effects [61-63].

The most common isocyanurates used in the coil industry are those based around aliphatic isocyanates, namely, hexamethylene diisocyanate (HMDI) and isophorone diisocyanate (IPDI). This is because aromatic diisocyanates have very poor weathering characteristics. The two mentioned can be used to achieve better formability or better durability. The choice is down to the formulator and the parameters that he needs to satisfy.

Figure 5.10 depicts an alternative road to polyurethanes. In order to avoid, because of its toxicity, free isocyanate during the course of manufacture, the synthesis is carried out by reaction of a polymeric carbonate ester with an amine.

Water dispersible polyurethanes (PUDs) are accessible by introduction of hydrophilic moieties, e.g. hydroxylated polyether side-chains into the PU backbone. There are 1- or 2-package paints available.

5.5.7 Acrylic resins [64]

Though acrylic resins are not widely used in coil coatings nowadays, their versatility, especially as waterborne dispersions, and their relatively low price makes them a viable alternative for future coatings.

Acrylic resins are commercially manufactured from acrylic or methacrylic acid esters, by means of catalysed radical polymerisation in solution or emulsion. The versatility of the ester building block, and the capability of copolymerising with other monomers, such

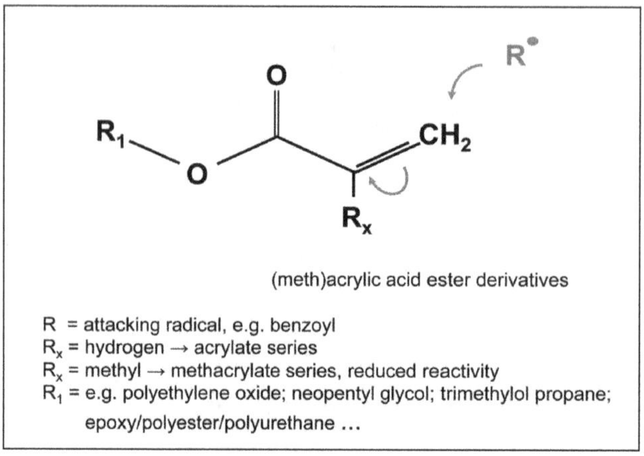

Figure 5.11: Acrylic and methacrylic ester derivatives as precursors for acrylic resins through a radical polymerisation

as vinyl acetate, styrene, etc., allows a large variety of resins to be obtained. At present, the worldwide annual production is reported above 1 Mt/a.

As an alternative to the production route from petrochemical precursors, acrylic monomers can also be obtained from 2-hydroxy propanoic acid (lactic acid) and its 3-hydroxy isomer.

Acrylics from saturated esters, styrene or vinyl toluene show a thermoplastic behaviour. OH functionalised monomers are more useful for coil coating applications, because they allow onward reactions of the polymeric strands with other functional binders, hence resulting in duroplastic (thermosetting) resins and films. Properties of the resin formulations and the resulting films can be tailored on purpose, including their rheology, glass transition temperature, melting point and mechanical features. Figure 5.11 shows the versatile building block of an acrylic resin: Acrylic or methacrylic esters that can be widely diversified in their ester function. Polymerisation takes place via radical mechanisms, the attack directed at the double bond. Up to date thermal processes are catalysed and controlled by atom, chain or group transfer reactions that allow step-by-step growth. Another option is the use of sterically hindered nitroxide radicals (otherwise used as light stabilisers), to achieve a temporary reversible blocking of the growing polymer chains. The radical mechanism can also be triggered by UV irradiation. Other unsaturated monomers like styrene, vinyl acetate, etc. can be copolymerised statistically or as blocks.

5.5.8 Binder resins: Comparison and uses

Table 5.2 summarises major performance features and uses of binder chemistries in coil coating paints [65].

Table 5.2: Common binder chemistries in coil coating paints Source: ECCA

Coating	Gauge [µm]	Characteristics	Limits	End uses
Polyester	25	Flexibility, thermal stability, outdoor durability resistance to household agents, detergents, humidity	UV stability requires protective foil for processing	Building, caravans, mobile homes appliance ("cold"), lighting
PVdF	25 to 30	Outdoor durability, chemical resistance (aliphatic hydrocarbons, fats)	Abrasion resistance	Building (wall cladding)
PVC plastisol	200	Formability, weatherability, corrosion resistance	Temperature limit 80 °C	Architecture (out & indoor), furniture, ducting
Polyurethane	25	Formability, UV and thermal stability, non-yellowing		Appliance ("hot" & dishwasher), external cladding

5.6 Coil coating paint statistics [66]

The official statistics of the European Coil Coating Association (ECCA) give a total volume of 193,000 t of paints sold in 2012. Of these paints, 112.6 kt were topcoats, 47.1 kt primers and 33.7 kt backing coats. The market has recovered from the previous recession, stabilising at a level just below the 2008 maximum.

Table 5.3: Coil coating paint sales in Europe 2012 Source: ECCA

Total volume 2012	
112.6 kt Topcoats	
47.1 kt Primers	
33.7 kt Backing coats	
Paint system	**% share**
Polyester	72.7
Polyurethane	11.7
PVC plastisol	10.9
PVdF	3.0
others	1.8
Paint demand total:	100.0
hereof: topcoats	58.2
primers	24.3
backing coats	17.4

The share split for the different binder chemistries considers the distribution of tonnage of paints sold. PVC plastisols are applied in about ten times the thickness that is common for all other paint systems. When correcting for this effect, polyester systems are used on more than 80 % of the surface area coated, polyurethane paints are second on almost 13 %, and PVdF on the third rank with over 3 %.

5.7 Roles and classes of pigments

5.7.1 Introduction

Pigments are classified according to the task they fulfil in a coating. Pigments are employed for corrosion protection, for structural and mechanical purposes, as colorants and for throwing power control (reflectivity, brilliance), and as extenders.

Corrosion protection pigments require a certain solubility to be effective, however must not be hygroscopic. They may act according to different principles, i.e. actively interfere with corrosion processes, form physical barriers, or corrode away instead of the base metal they protect (cf. also Chapter 2.2).

The coloration of a coating is brought about by pigments that may be solid coloured, or achromatic (white, black). The aspect of a film is furthermore influenced by the opacity and reflectivity, or light scattering properties (as with metallic or mother-of-pearl effect paints).

Fillers are primarily used as volume extenders, however they may have additional benefits like reinforcement or improvement of chemical properties.

5.7.2 Anticorrosive pigments [67, 68]

5.7.2.1 Overview

Active corrosion inhibitor pigments act along three mechanical pathways: Strong oxidants like chromates have been used to passivate bare metal at a failure spot, consuming electrons that are released by metal dissolution. Direct replacements for the toxic chromates are being sought after.

Growth of passive layers or band gap widening in mixed metal oxides may disable the transition of electrons through a passivation layer and thereby effectively suppress oxygen reduction.

Ion-exchange pigments like calcium-modified silicates release calcium ions (Ca^{2+}) in the presence of acids, thereby reducing the corrosivity of the electrolyte, and intercepting corrosive ions like chlorides or sulphates.

Metallic pigments with a lower corrosion potential (Zn, Al, Mg…) are used as sacrificial anodes that corrode in place of the nobler base metal. Part of the protective effect is attributed to the cementation of insoluble corrosion products that may fill pores and scratches. In particular, zinc pigmented primers have been used as anti-corrosion barrier for various artisan and industrial coating uses. In a coil coating application, zinc-rich corrosion protection primers (CPP) are applied on sheet for the car manufacture. Improved corrosion protection is achieved in critical areas of the car body, like seams, box sections or flanges [69, 70].

Barrier pigments from e.g. chemically inert aluminium, glass flakes, mica or micaceous iron oxide are described to extend the diffusion path for corrosive species through a coating. They may also play a part in structurally reinforcing the binder matrix and interfering with its water permeability.

5.7.2.2 Active pigments – lead and chromate pigments [71]

Historically, lead (in particular red lead, lead tetroxide Pb_3O_4) and chromate compounds have been the most important anticorrosive pigments. Both classes, however, are subject to strict environmental control and reduction measures due to their high toxicity and carcinogenity. Lead compounds already disappeared from the coil market, while chromate pigments still await their complete replacement.

Chromate, as a strong oxidant with E_{pot} = 1.195 V, is slowly released from the pigments and made available to passivate metal surfaces by formation of a dense, almost impermeable mixed oxide layer. Strontium chromate ($SrCrO_4$), zinc potassium chromate (zinc yellow, $K_2Zn_4(OH)_2(CrO_4)_4(H_2O)_2$, and zinc tetrahydroxy chromate, $Zn_5(OH)_8(CrO_4)$, have been particularly important in industrial primers.

5.7.2.3 Active pigments – phosphate pigments [72]

Zinc phosphate, either of di- or tetrahydrate type, $Zn_3(PO_4)_2$ $(H_2O)_x$, and modified orthophosphates (basic Zn phosphate or Zn Al phosphate) are used as a replacement for chromates. Unlike these, however, the protection mechanism is merely attributed to the formation of dense precipitate films of phosphate or basic phosphate that are able to block the access of oxygen to the corrosion site. Phosphate/borate and phosphate/silicate types of pigments are also used, which display a higher release rate of phosphate, enabling faster film formation on failure sites. The protective mechanism of polyphosphates like acidic aluminium tripolyphosphate, $AlH_2(P_3O_{10})(H_2O)_2$, and its modifications with Zn, Ca or Sr, is explained by their chelating nature and their ability to release high amounts of phosphate upon hydrolytic depolymerisation.

Polyphosphate pigments are also known in synergistic combinations with organic compounds like organic acids, amines, or certain organic oxidants. More recent studies have involved the fine milling of phosphate pigments to a size of 100 to 200 nm, used in a 60 μm model epoxy primer coating. Enhanced barrier and anticorrosive properties were obtained [73]. There is a fair chance that this improved performance will also be confirmed when applying at a low dry film thickness typical for coil coating.

5.7.2.4 *Other active anticorrosive pigments* [74]

Basic zinc molybdo phosphates have an improved anticorrosive activity since the oxidative power of molybdate is just short of chromates. A similar mechanism applies for blends from molybdate and zinc oxide, zinc phosphate or calcium carbonate.

Zinc hydroxyphosphite is described to form passivating iron phosphite and phosphate layers on steel substrates, as well as zinc oxide and soaps that serve as a physical barrier. Other active inhibitors comprise borates, silicoborates and zinc cyanamide.

Calcium doped silica preparations have long been invented as a class of anticorrosive pigments. They are formed by the reaction of lime with the hydroxy groups that are present on the porous surface of silica particles. Their anticorrosive effect is attributed to a multiple mechanism. Upon corrosive attack, Ca^{2+} ions are released in exchange for small, aggressive cations (H^+, Na^+) from the surrounding electrolyte, thereby reducing the corrosive effect. Also, accompanying anions (Cl^-, SO_4^{2-}) are immobilised through precipitation by the Ca^{2+}. Silica moreover interferes with the cathodic branch of the corrosion reactions, by consuming the hydroxyl ions (OH^-) that are generated in the course of the oxygen reduction. Thus, this reaction is not subdued itself, but its degrading effect through the increased alkalinity at the surface diminished (see Chapter 2.2). Other dope additives studied comprise Mg^{2+} and $Ce^{3+/4+}$. Particular undisclosed features in recently developed Ca doped pigment preparations, concerning the load type and amount, as well as the aggregation, are reported to provide anticorrosive effects on HDG substrate comparable to the performance of conventional strontium chromate pigmentation [75]. Figure 5.12 shows a microscopic image of the material.

Figure 5.12: Micrograph of a porous silica particle Source: Grace

Also mineral or synthetically generated particles like hydrotalcites or zeolites can be designed as carriers to host e.g. anticorrosive ions like molybdates or vanadates, or calcium ions as interceptors to corrosive species like chloride. Such carrier concepts are considered to be particularly useful, because they enable retarded release of the active agents that thus remain available for longer periods, instead of quickly bleeding out of the coating, depriving it of its properties. The release might even be controlled by the conditions of the surrounding medium, e.g. pH value, temperature or the presence of (corrosive) counter ions, so that the active ions are only available on demand.

5.7.3 *Barrier and sacrificial pigments* [76]

Barrier pigments must be entirely insoluble and inert to exert optimum performance.

Lamellar pigments such as micaceous iron oxide (α-Fe_2O_3, haematite) are supposed to form a barrier of aligned flat and thin particles, thereby reducing the permeability of the paint film by extending the diffusion path for any penetrating corrosive species. Moreover, ultraviolet (UV) absorption, film reinforcement and increased intercoat adhesion are reported.

Non-leafing aluminium flakes and zinc flakes are also used as barrier pigments in solvent-borne paints.

Spherical zinc dust particles are used in so-called zinc-rich coatings at levels of 70 % or more. When densely packed in the film, having electrical contact with each other and the ferrous substrate, the zinc particles corrode preferentially, leaving the base metal intact. Corrosion products are deposited by way of cementation and block pores and crevices in the coating. Encapsulation of the Zn particles with nanoscale titanium dioxide layers, as has been reported for other applications [77], might also render coil coatings that are particularly resistant to the formation of white rust. Magnesium particles are studied for a similar use on Al substrates. Safety concerns related to the flammability of the Mg pigment may be comforted again by thin surface oxide layers applied on the particles.

5.7.4 Colouring and extender pigments [78]

Colouring pigments can be classified in various ways, according to their use, principle mechanism, chromatic features, or their chemical nature and origin. Inorganic pigments have been developed employing a wide spectrum of chemical and mineral compositions. A classification chart is displayed in Figure 5.13.

Titanium dioxide, TiO_2, in its rutile modification, is by far the most important white pigment. A phosphate modified titanium dioxide is known to have additional anticorrosive properties. Also catalytic effects are reported that help in the destruction and removal of organic scale that adheres to a coating.

Baryte, barium sulphate, "Blanc fixe", or chalks are used to a lesser extent, since they have much less hiding power. On the other hand, they improve the acid and chemical resistance.

Carbon blacks and iron oxide blacks are the most common black pigments. Being efficient UV absorbers, they impart excellent weatherability upon outdoor exposure.

To provide the desired colour range, organic and inorganic pigments are used alike, the choice being made considering the chroma, opacity, dispersability and overall economy of the pig-

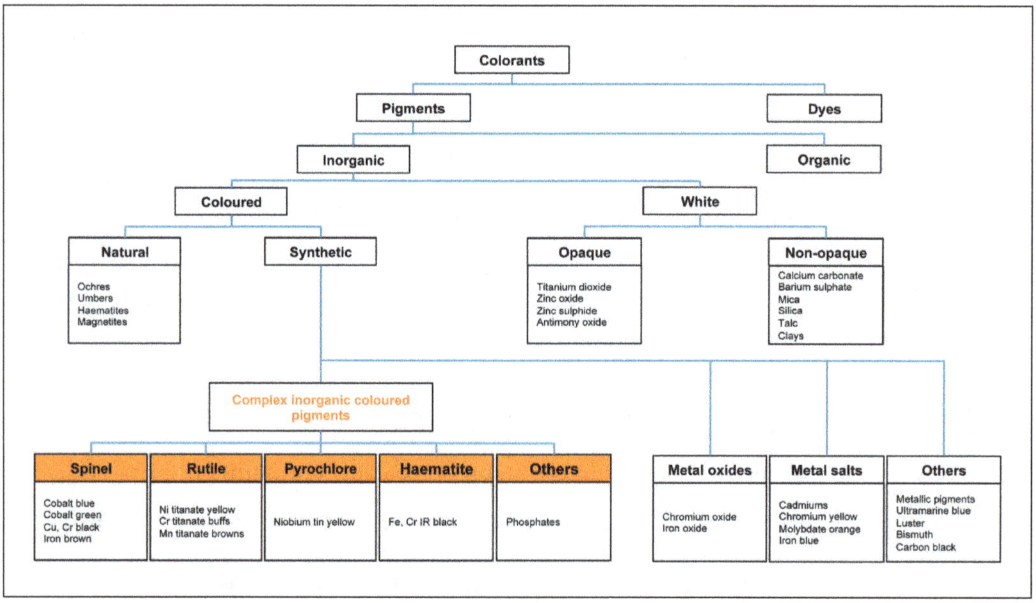

Figure 5.13: The family tree of colouring pigment classes *Illustration according to Ryan* [73]

ments. Also here, chromate and lead pigments, frequently used in the past for their excellent colour intensity, are being replaced by less toxic materials [79, 80]. Red tones are obtained with iron red oxides or quinacridones, yellow to orange shades with yellow ferrites (hydrated iron oxide), bismuth vanadates, titanium yellows, tin-niobium or tin-zinc oxides (Figure 5.14), or benzimidazolones. The green-to-blue range is covered by chromium oxide green and phthalocyanines. Where appropriate, pigment blends are formulated to adjust colour tones.

Extender pigments like carbonates, silicates, talc, muscovite mica etc. are often added to give extra volume, however they may also exhibit additional benefits. For instance, wollastonite (calcium silicate) acts as a reinforcer and pH buffer for paint films, along with its effect to reduce permeability via improved pigment/binder interfaces. The doping of calcium carbonate with magnesium (dolomite, $[Ca,Mg]CO_3$) leads to improved acid resistance, kaolin (aluminium magnesium silicate) adds to the mechanical resistance. Scratch resistance of a coating is supported by addition of nanoparticulate compounds, e.g. alumina or silica.

5.8 Auxiliaries

Apart from solvent carrier, binder and pigmentation to obtain colouring and corrosion protection, organic coating chemicals contain small amounts of a variety of additives for specific purposes [81]. These additives may comprise

- Crosslinking catalysts
- Hydrophobicisers, plasticisers, matting agents
- UV absorbers, e.g. "hindered amine light stabiliser" radical quenchers (HALS)
- Dry lubricants and slip additives
- Surfactants for wetting, spreading, dispersing, defoaming, degassing
- Anti-microbial agents (general-purpose or specific interceptors)
- Additives for fire retardant or fire resistant coatings

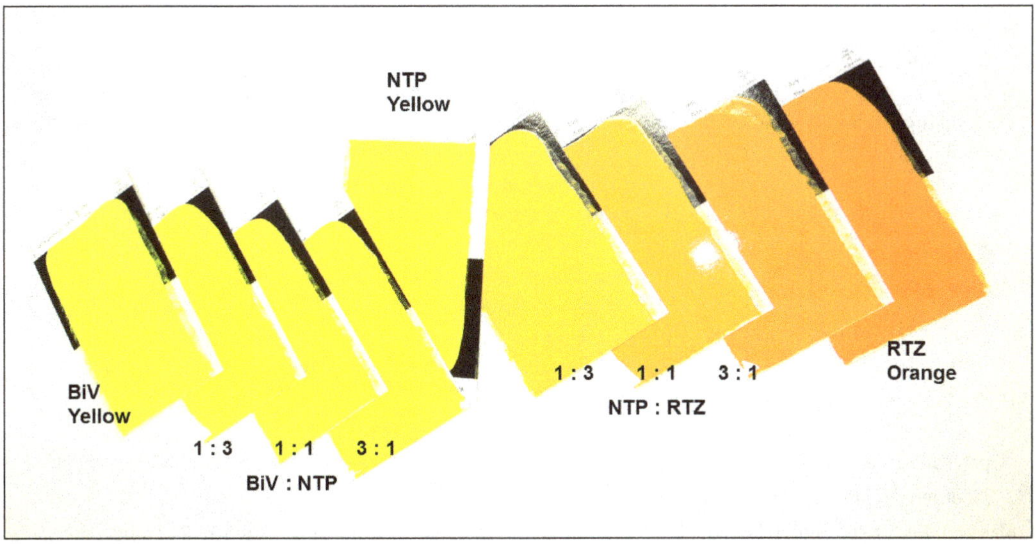

Figure 5.14: Formulation of paints with yellow to orange colours, using non-toxic replacements for lead chromate. BiV: Bismuth vanadate; NTP: Niobium tin pyrochlore, $Sn_2Nb_2O_7$; RTZ: Rutile tin zinc

Source: Shepherd Colors

5.9 Literature

[1] Manea, M., in: Sander, J., Kirmaier, L., Manea, M., Shchukin, D., Skorb, E., Anticorrosive Coatings, Vincentz Network, Hannover 2010, p. 37

[2] Goldschmidt, A., Streitberger, H.-J., Basic of Coating Technology, 2nd ed., BASF Coatings AG, Münster 2007, pp. 127 ff

[3] Meuthen, B., Jandel, A.-S., Coil Coating, 2nd ed., Vieweg, Wiesbaden 2008, p. 71

[4] anon., Dampfdruck, molare Verdampfungsenthalpie und Raoultsches Gesetz (Vapour pressure, molar evaporation enthalpy ...), Univ. Jena 2007 (http://www.ipc.uni-jena.de/downloads/Lehre/PCII/Praktikum/PCII_06_Verdampfungsenthalpie.pdf; 16-07-13; 22:03 h

[5] Romey, E., Morlock, A., Wasserlack oder High Solid? Schritt für Schritt zum richtigen Lacksystem (Waterborne paint or high-solid?...), JOT 49, 2009, pp. 32 ff

[6] anon., Evaporation rates of solvents, Siegwerk Switzerland, Bargen, CH, 2008 (http://www.siegwerk.com/fileadmin/user_upload/cc/Data_Sheets/TM/Verdunstungsgeschwindigkeit_e.pdf; 16-07-13; 18:11 h

[7] anon., 1-butanol - Compound Summary, National Center for Biotechnology Information, U.S. National Library of Medicine, Bethesda MD, USA 2013 (http://pubchem.ncbi.nlm.nih.gov/summary/summary.cgi?cid=263; 16-07-13; 17:18 h

[8] anon., Butyl glycol, Techn. Leaflet, BASF, Ludwigshafen 2008 (http://www.solvents.basf.com/portal/load/fid226758/ButylGlycol_e_04_08.pdf; 16-07-13; 22:08 h

[9] anon., n-butoxyethanol - Compound Summary, National Center for Biotechnology Information, U.S. National Library of Medicine, Bethesda MD, USA 2013 (http://pubchem.ncbi.nlm.nih.gov/summary/summary.cgi?cid=8133; 16-07-13; 17:16 h

[10] anon., 3-xylene - Compound Summary, National Center for Biotechnology Information, U.S. National Library of Medicine, Bethesda MD, USA 2013 (http://pubchem.ncbi.nlm.nih.gov/summary/summary.cgi?cid=7929; 16-07-13; 17:19 h

[11] anon., Solvesso series aromatic hydrocarbons, Multisol Ltd., Nantwich, UK, 2013 (http://www.multisolgroup.com/Solvesso-Series-Typical-Properties.pdf; 16-07-13, 14:14 h

[12] anon., Petrol 24534, Safety Data Sheet, Sigma Aldrich Co.LLC, Gillingham, UK, 2012 (http://www.sigmaaldrich.com/catalog/product/sial/24534?lang=en&region=GB; 16-07-13, 17:04 h

[13] Weast, R.C., Astle, M.J. (eds.), Handbook of Chemistry and Physics, 60th ed., CRC Press, Boca Raton FL, USA, 1980, pp. C-81 ff; C-727 ff; E-386; F-45 ff

[14] Sander, J., Kirmaier, L., Manea, M., Shchukin, D., Skorb, E., Anticorrosive Coatings, Vincentz Network, Hannover 2010, pp. 92 f

[15] Sander, J. et al. [14], p. 99

[16] Goldschmidt, A., Streitberger, H.-J. [2], pp. 337 ff

[17] Casagrande, M., Untersuchung der Ordnung und Diffusion von Polymerdispersionen bei ihrer Verfilmung (Investigation of the order and diffusion of polymer dispersions ..., Diss., Joh. Gutenberg Univ., Mainz 2000, pp. 8 f; http://archimed.uni-mainz.de/pub/2001/0061/diss.pdf; 14-08-13, 12:44 h

[18] Jahr, T., Untersuchung der Filmbildung aus Polymerdispersionen mit Hilfe der forcierten Rayleighstreuung (Investigation of the film formation from polymer dispersions ...), Diss., Joh.-Gutenberg-Univ., Mainz 2002, p. 14 ff; httpd-nb.info96420774534; 06-05-13, 17:55 h

[19] anon., Trocknung von Latex-Filmen (Drying of latex films), TU Clausthal 2013; www.pc.tu-clausthal.de/forschung/ak-johannsmann/trocknung-von-latex-filmen/; 06-05-13, 17:48 h

[20] anon., Stages of film formation - Film formation from latex dispersions, Univ. Freiburg 2013; www.colloids.uni-freiburg.de/copy_of_Methoden/filmform.htm?set_language=en; 06-05-13, 17:35 h

[21] Sander, J. et al. [14], pp. 100 f

[22] Giannakopoulos, I., The effect of cross-linker concentration on the fracture toughness of free-films of polyester based, HMMM cross-linked coil-coatings, European Coatings Conference, Proc., Vincentz 2013

[23] Meuthen, B., Jandel, A.-S. [3], p. 130

[24] Kittel, H., Streitberger, H.-J. (ed.), Lehrbuch der Lacke und Beschichtungen (Coursebook of Paints and Coatings), Vol. 6, 2nd ed., Hirzel, Stuttgart 2008, pp. 286 ff

[25] Lowe, C., Fully cured or just a little better, ECCA Autumn Congress, Proc., European Coil Coating Association, Brussels 2008

[26] Brown, N., Radiation Curing Systems in Coil Coating, ECCA General Meeting, Proc., European Coil Coating Association, Brussels 2006

[27] Heylen, M., Franzolin, G., Radiation Curable Coil Coatings: Innovative and Feasible Solutions for the Steel and Aluminium Industry, ECCA, Autumn Congress, Proc., European Coil Coating Association, Brussels 2005

[28] Sander, J., Koch, M., Picot, P., Réticulation UV - Une nouvelle approche pour le „coil-coating" (UV Curing - A New Chemical Approach for Coil Coating), Centre d'Actualisation des Connaissances et de l'Etude des Matériaux Industriels (Cacemi), Journée spécialisée, Proc., Paris 2002

[29] Lowe, C., WeatherEBility - 15 Years Exterior Weathering Experience of Some EB Curable Coatings, ECCA Autumn Congress, Proc., European Coil Coating Association, Brussels 2009

[30] Sander, J., Up To Scratch With Novacoat 1000 UV, Steel & Aluminium Newsletter, Henkel KGaA, Düsseldorf 2007

[31] anon., Edler Touch für Edelstahl - Neue Oberflächenbehandlungen erhalten die elegante Optik des Edelstahls (Noble Touch for Stainless Steel - New Surface Treatments Preserve the Elegant Aspect...), Stahlmarkt 04, 2007, pp. 50 f; cit. re. Meuthen, Jandel [3], p. 136

[32] Sander et al. [14], pp. 93 f

[33] Kloppers, P., A New Powder Application System - EMB: Electro Magnetic Brush, 8th International NIR-Symposium, Proc., adphos GmbH, Bruckmühl 2012

[34] Kloppers, P., EMB status update, ECCA Autumn Congress, Proc., European Coil Coating Association, Brussels 2009

[35] Gedan-Smolka, M., Strohbeck, U., Demmler, M., Cudazzo, M., Vogelsang, H., Tuschla, M., Ressourcenschonendes Pulver-Precoating-Verfahren eröffnet neue Märkte für vorbeschichtetes Blech (Resources saving powder pre-coating process ...), 77. Lacktagung, Proc., GDCh (German Chem. Soc.), Frankfurt a.M. 2012

[36] Ritterbach, A., Schnelle Pulverlack-Aushärtung mit Induktion: Coil-Coating im Kleinformat (Fast Curing of Powder Coats by Induction...), JOT 47, 2007, pp. 12 ff

[37] Bär, K.K.O., New Opportunities, Added Value and Lower Cost Coil Coating with Strong Reduced Emissions, ECCA Autumn Congress Proc., European Coil Coating Association, Brussels 2012

[38] Bielefeld, F.W., Sander, J., The profitable choice - From small-batch post painting to compact coil coating, Europ. Coatings Conf. Market Day Coil and Can Coatings, Proc., Vincentz Network, Hannover 2011

[39] Manea, M. [1], pp. 38 ff

[40] Manea, M. [1], pp. 46 f

[41] Buszard, D. L., Commercial Plasticisers, in: Titow, W. V. (ed.), PVC Technology, 4th ed., Elsevier Appl. Sci. Publ., Barking, GB 1984, pp. 147 ff

[42] Wypych, G., PVC Degradation & Stabilization, ChemTech Publ., Toronto ONT, Canada 2008, p. 122

[43] Owen, E.D. in: Owen, E.D. (ed.), Degradation and Stabilisation of PVC, Elsevier, London 1984, Chapter 5, pp. 215 ff

[44] Titow, W. V., Elementary Principles of PVC Formulation, in: Titow, W.V. (ed.), PVC Technology, 4th ed., Elsevier Appl. Sci. Publ., Barking, GB 1984 pp. 79 ff

[45] anon., EPA Retreats on 2 Chemical Rules, Durability + Design, Technology Publ. Co., Pittsburgh PA, USA 2013; http://www.durabilityanddesign.com/news/?fuseaction=view&id=10255&nl_versionid=3418, 20-09-13, 09:55 h

[46] Meuthen, B., Jandel, A.-S. [3], p. 62

[47] Swartz, N.A., Wood, K.A., Clare, T.S., Characterizing and improving performance properties of thin solid films produced by weatherable water-borne colloidal suspensions on bronze substrates, Progr.Org.Ctgs. 75, 2008, pp. 215 ff

[48] anon., The short and the long story of epoxy resins, Royal Dutch Shell (ed.), Amsterdam 1992, p. 14

[49] anon., EPA Moves Step Closer to Possible Regulatory Action on Bisphenol A, J. Architect. Coatgs., 2011

[50] anon. [45]

[51] Manea, M., High Solid Binders, Vincentz Network, Hannover 2008, p. 98

[52] anon., [48], p.63

[53] anon., [48], pp. 82 ff

[54] Manea, M. [50], pp. 85 ff

[55] Wicks, Z.W., Jones, F.N., Pappas, S.P., Wicks, D.A., Organic Coatings: Science and Technology, 3rd ed. 2, Wiley, Hoboken NJ, USA 2007, pp. 200 ff

[56] Manea, M. [51], pp. 151 ff

[57] Schmitthenner, M., Basic resins for coil coatings, ECJ, 1998, p. 618

[58] Maxted, J., Hitting the Target: Advances in Coating Curing Monitoring, ECCA Autumn Congress, Proc., European Coil Coating Association, Brussels 2009

[59] Manea, M. [1], pp. 55 ff

[60] Manea, M. [51], p. 48

[61] Hoch, M., Organotin compounds in the environment - an overview, Appl. Geochemistry 16, 2001, pp. 719 ff

[62] anon., Product Safety Assessment: METATIN™ Dibutyltin Catalysts, Leaflet, The Dow Chemical Co., Midland MI, USA 2011; http://msdssearch.dow.com/PublishedLiteratureDOWCOM/dh_0886/0901b80380886a81.pdf?filepath=productsafety/pdfs/noreg/233-00882.pd.&fromPage=GetDoc (27-07-13; 16:16 h)

[63] Richter, F., Fleck, O., Halpaap, R., Hecking, A., Grahl, M., Iovkova-Berends, L., Bradtmöller, G., Zöller, T., Jurkschat, K., Thermolatente Katalysatoren für die Urethanbildung (Thermo latent catalysts for urethane formation), 77. Lacktagung Bremerhaven, Proc., Ges. Dtsch. Chem. (German Chem. Soc.), Frankfurt a.M. 2012

[64] Manea, M. [1], pp. 53 ff

[65] anon., ECCA Academy 2013; www.prepaintedmetalacademy.eu

[66] anon., [65]

[67] Kirmaier, L. in: Sander, J. et al. [14], pp. 73 f

[68] Sander, J., Corrosion Protection Coatings, European Coatings Seminars, Amsterdam 2012, Proc., Vincentz Network, Hannover 2012

[69] Meuthen, B., Jandel, A.-S. [3], pp. 75 ff

[70] Schinzel, M., Advanced Corrosion Protection of Automotive Body Sheet - A Challenge for Coil Coating, ECCA Autumn Congress, Proc., European Coil Coating Association, Brussels 2009

[71] Kirmaier, L. [67], pp. 74 ff

[72] Kirmaier, L. [67], pp. 77 ff

[73] Entenmann, M., Greisinger, H., Maurer, R., Schauer, T., Corrosion protection with nanoscale anticorrosive pigments in coatings, ECJ 06/2011, pp. 29 ff

[74] Kirmaier, L. [67], pp. 80 ff

[75] Fletcher, T.E., Kent, D.J., New developments towards the replacement of chromates in coil coating primers, ECCA Autumn Congress, Proc., European Coil Coating Association, Brussels 2012

[76] Kirmaier, L. [67], pp. 85 f

[77] Götz, H., Zinklack mit Nanoeffekt (Zinc paint with nano effect), JOT 50, 2010, pp. 32 f

[78] Kirmaier, L. [67], pp. 87 ff

[79] Ryan, M., New Ultra-High Performance Inorganic Pigments to Replace Lead Chromate, ECCA Autumn Congress, Proc., European Coil Coating Association, Brussels 2012

[80] Korona, E., Lead free pigments for coil coating, ECCA Autumn Congress, Proc., European Coil Coating Association, Brussels 2010

[81] Sander, J. et al. [14], pp. 90 f

6 Coil coating line equipment

6.1 Cleaning and pretreatment installations

6.1.1 Cleaning

The features of the cleaning operation have been described in detail in Chapter 3.

Most cleaning sections in coil coating lines are constructed for spray cleaning. Several separate cleaning and rinse stages form a tunnel that is passed by the moving strip. The hoods of the individual compartments are usually designed as vapour-tight, but hydraulically removable lids, thereby enabling easier maintenance. On the entrance and exit of each individual cleaner and rinse stages, squeeze rolls are mounted that support the strip position, form a slot against overspray and aerosols, and minimise the liquid that is dragged out by the moving strip into the following stage. As mentioned earlier, the loss of liquid can be as low as 2 to 5 ml/m^2.

The rinses are operated as a counterflow cascade, to ensure effective rinsing at minimised water consumption (cf. Chapters 3.4.2, 3.4.3). The current water consumption in spray processes in a coil coating line averages to 12.7 l/m^2 of precoated strip (cf. Chapter 6.4.1.1, Table 6.2) [1].

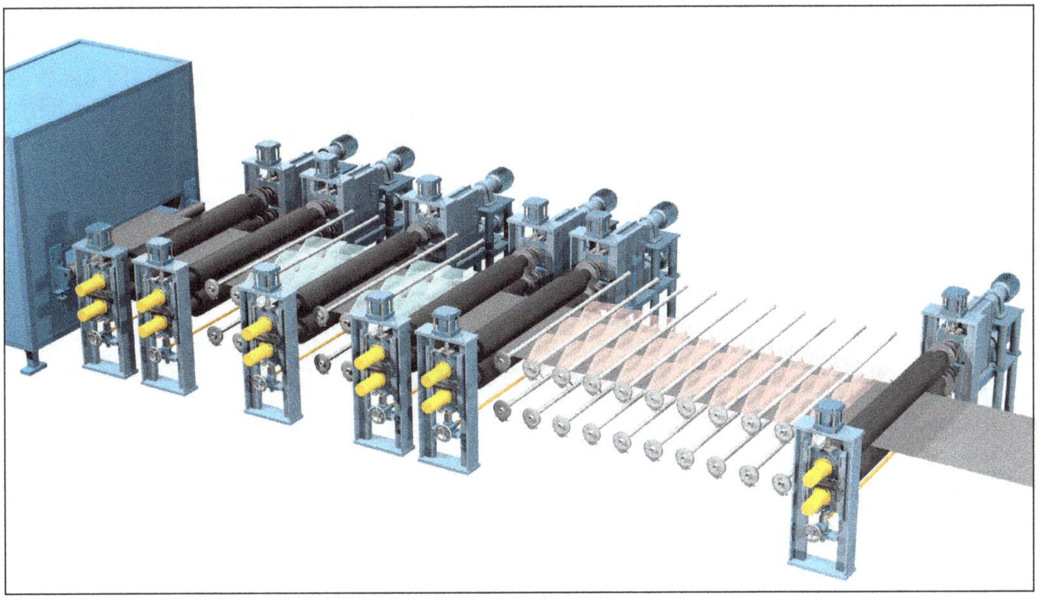

Figure 6.1: Explosion view of a modern cleaning stage with spray section and cascaded rinses in a coil coating line

Source: SMS Siemag

Figure 6.2: Spray cleaning section from the operator side. The tank is mounted under the tunnel, and provides easy access for control and maintenance Source: Globus

In each stage, a number of individual spray bar tubes, studded with nozzles, are mounted perpendicularly across, to cover the strip from its top and bottom faces. These spray bars are fitted with fast coupling devices to allow their quick removal and exchange for maintenance.

The number, size (flow capacity) and shape of the nozzles are selected to provide the correct supply of cleaner solution at an appropriate, effective pressure. Most commonly, flat fan nozzles are used that ensure a sharp, flat jet impacting on the strip surface. The jet fan usually opens in a 60° angle. In particular, tongue nozzles are chosen for their robustness and non-clogging operation. The nozzles are mounted with a small incline in two spatial directions so that the jets slightly point against the strip travel direction, and their fan centre lines are in offset. This enables optimum overlap of the spray pattern while avoiding the mutual annihilation of neighbouring jets [2, 3].

Thin-gauge and light-weight coil is also cleaned in immersion tanks or in combined dip/spray cleaning compartments. To immerse the coil into the bath, dip rolls are used. A number of dip rolls can be mounted in a way that the strip forms several loops in the bath, in order to extend the available treatment time. The speed of the travelling strip is usually sufficient to bring on the required exchange of reactants from the surface. To enhance the turbulence, the baths may be additionally equipped with injection nozzles.

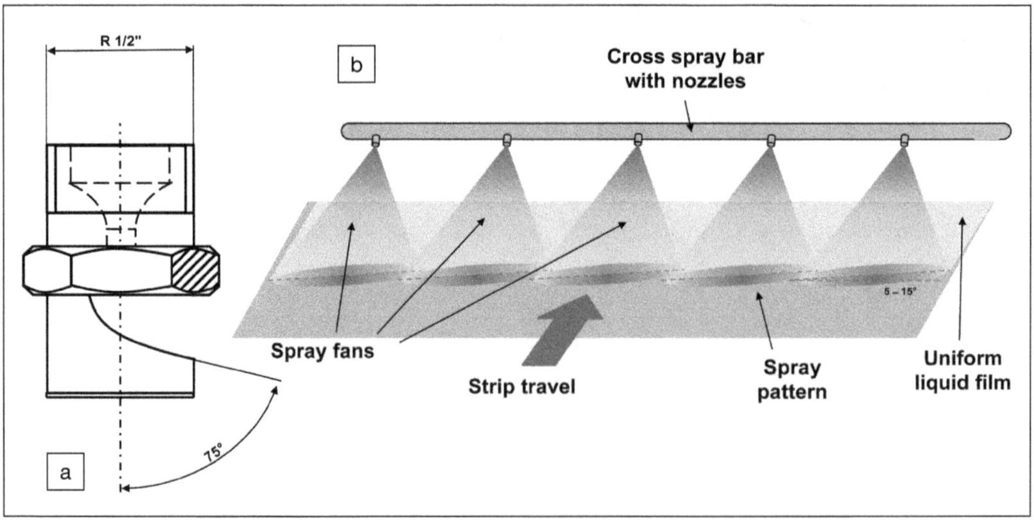

Figure 6.3: a) Technical drawing of a tongue nozzle commonly used in spray cleaning; b) offset arrangement of spray nozzles ensures optimum distribution of the cleaner on the moving strip surface Illustration acc. to Lechler [1, 2]

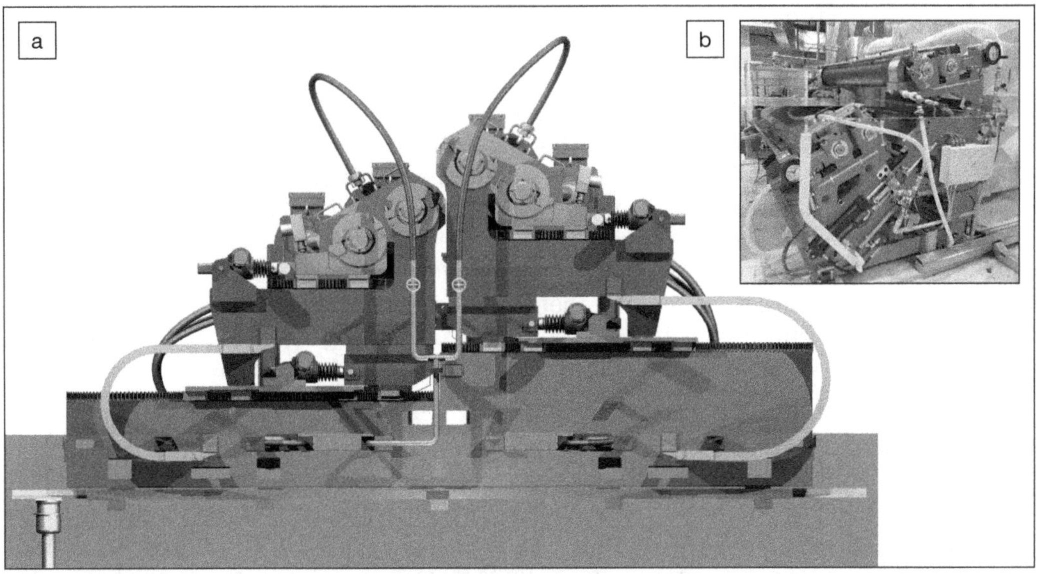

Figure 6.4: a) Drawing of a vertical chemcoater; b) horizontal chemcoater in standby mode
Both 2 · 2 roller machines are designed for operation against the strip tension Source: a) SMS Siemag, b) Globus

6.1.2 Pretreatment application

Immersion or spray techniques are also common for the application of conventional pretreatments. In principle, the same considerations regarding the efficient use of treatment and rinse fluids apply as discussed for the cleaning process (cf. previous chapter). Hence, also the same technical solutions have proven their suitability. More information about installation features directly related to the chemical process can be found in Chapter 4.6.

In more recently erected coil coating lines, no-rinse or dry-in-place pretreatments are commonly used. They are applied by especially adapted roll-coaters (chem-coaters). The water consumption by no-rinse technology is reduced by a factor of 7.5 as compared to conventional spray technique (cf. Chapter 6.4.1.1, Table 6.2), going along also with a low waste water generation. Efficient chemical usage and a reliable quality of the application are further benefits that can be gained through no-rinse processes.

Figure 6.4a and 6.4b show different chem-coater layouts for vertical and horizontal installation (see also Chapter 1, Figure 1.6). More details on the design and operation of roll-coaters are discussed in Chapter 6.4 below.

Prior to the painting stages, the strip needs to be thoroughly dried. Short hot air convection or floatation ovens are sufficient to evaporate the water (Figure 6.5). Usually the strip needs to be heated to only 80 °C to obtain the desired effect.

Figure 6.5: Hot air drier mounted after a chem-coater Source: Globus

6.1.3 Advances in plasma technology for cleaning and pretreatment

Plasma cleaning is being investigated as a potential means to avoid the management of cleaning fluids and wastes. Plasma, e.g. gas ionised in an open arc discharge flame, when accelerated against a soiled solid surface, transfers chemical bonds in contaminant organic molecules into an excited state, thereby breaking up the bonds and finally destroying and removing the molecule. The technique can be operated under normal atmospheric pressure, doing away with the vacuum equipment that was formerly required. With conventional, dot-shaped plasma sources, an array of them is required that covers the width of the strip and also a certain length of it. In an aluminium coil coating operation for narrow strip (sunblinds) this was realised by arranging 24 plasma canons in eight rows radially around a support roll for either side of the strip. The line is capable of a speed up to 40 m/min [4].

Plasma sources with a linear slit outlet shall provide a more efficient means. Currently, slit lengths of 350 mm are realised. Wider webs can be covered by combination of several of such sources. Experiments were reported with different plasma gas compositions. In an argon/nitrogen plasma (Ar/N = 1.5 to 1.8 : 1), oil films of up to 1.0 g/m^2 were removed from Al sheets in a single pass, at a process speed up to 50 m/min. These data, though being still short of the conditions in a large coil coating installation, nonetheless show some promise for the technology. An effective flow rate of plasma gas is reported in the order of 100 to 160 l/min (170 to 270 Pa m^3/s) under standard conditions (273.15 K = 0 °C; 1.01325 bar) [5].

Plasma sources can also be used for the chemical conversion of metal surfaces (cf. Chapter 4.5.1). By introduction of so-called precursor substances into an electric arc or combustion plasma (combustion chemical vapour deposition, CCVD), thin layers of up to several 100 nm can be deposited on precleaned metal surfaces. Hence, a combination of two plasma stages, the first without, the second with precursor gas feeding, would constitute a feasible array for fluid-free conversion treatment. Reportedly, the obtainable layer thickness depends on the flow rate and composition of the precursor and carrier plasma, distance of the plasma source to the substrate and temperature. Organosilanes are used as common precursors rendering nanoscale silica layers. Appropriate conditions providing, the CCVD process renders layers that display a good wettability. Fundamental studies were reported on galvanised steel, regarding the parameters and obtainable performance of the deposited layers. To ensure good barrier properties, a necessary treatment time of ≥ 1 s was established. An additional treatment with functionalised silanes (e.g. 3-glycidyloxypropyl triethoxy silane) is required to obtain good adhesion of subsequent paint or adhesive coatings and good anticorrosive performance. The necessity of such additional treatment has prevented the industrial application of the process so far. A commercial application of the CCVD technology is reported on stainless steel strip [6–9].

6.2 Strip travel and planarity control

Over the recent years, improvements in the control of strip travel and planarity have been in the focus of development.

Strip levelling is necessary to eliminate unevenness in the strip that is caused by either irregular or regular rolling failures (errors in roll bite adjustment, chattering etc.), or irregular heating and cooling conditions and resulting strain. Waves or local variations in thickness correspond to a variation in strip length (Figure 6.6). Hence, when failures occur mostly on one side of the strip, or the strip is even rolled with a wedge-shaped cross-section, the thicker edge is also shorter, resulting in the strip forming a crescent along its length

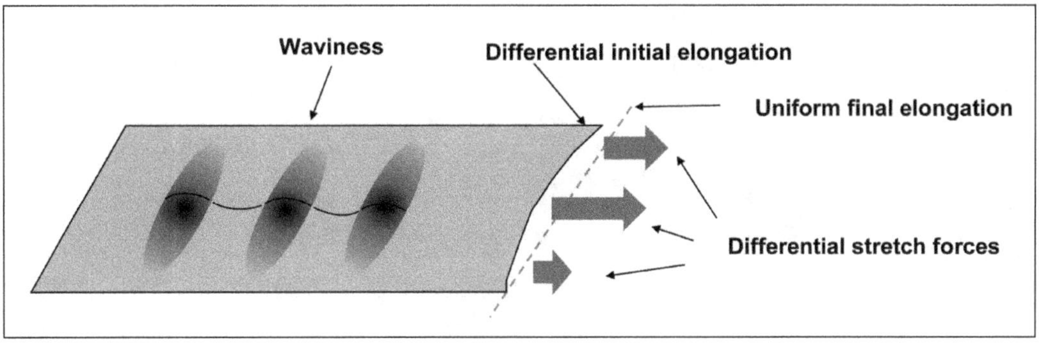

Figure 6.6: Schematic representation of waviness in a metal coil and its removal by differential stretching Illustration according to Koppers et al. [10]

(camber). Also, plastic deformation can occur along the strip (coilset) or across (crossbow) when bending around a deflector roll. To obtain a uniform elongation in all places, the strip must be plastically stretched. For proper levelling, also the elasticity of the metal has to be considered that leads to the material springing back into its original shape when the tension is released [10].

In most flat steel operations, this flatness correction is integrated in other processing stages like pickling, temper-rolling, annealing, or galvanising. However, in flat aluminium finishing, levelling machines are often part of aluminium coil coating lines.

Modern stretcher-leveller equipment has been put on service that allows a very tight control of the coil flatness. Figure 6.7 illustrates the principle setup. Figure 6.8a to 6.8d shows units in operation.

The stretcher-leveller is preceded and followed by a set of bridle rolls that build up the tension. In the stretcher-leveller, the coil winds through a staggered array of small-diameter rolls. The upper set of these rolls dips into the travel path of the strip so that the coil is partially wrapped around each roll, thereby bending and stretching the metal by nuances, and ensuring the strip is stretched flat. To reduce mechanical work (cold working) and wear, the number of these levelling rolls usually is chosen to be 3 or 5. Levelling roll positions

Figure 6.7: Drawing of a stretching-levelling unit Source: SMS Siemag

Figure 6.8: Modern stretching-levelling units with precise flatness control; a) tension bridle rolls; b) bridle drive with flexibly adjustable AC motors; c) tension-leveller with one crossbow and two flex sections; the penetration is adjusted by AC motors via screw jacks; d) levelling rolls built in individually driven cassettes in 4-high arrangement; the upper set of rolls can be adjusted in height to control the wrap angle of the strip around each roll that effects the stretching Source: a, d) BWG; b, c) Globus

and rotatory speeds are individually and flexibly adjusted. Similarly, individual drives are required for the bridle rolls, e.g. by AC motors with variable speed and transmission. The adjustment must be done by extremely small increments, to cope with the differential speed between the rolls without slipping and vibration. Modern equipment provides strip tensions up to 50 t. Automated processing uses precise tension, flatness or elongation control signals in a closed-loop control, thus allowing the deviation of flatness and strip elongation to be kept within extremely narrow limits, i.e. no more than ± 0.01 % [11, 12]. Flexible stretching-levelling technology was first invented in the late 1960s, and has been enhanced and augmented with auxiliary equipment and computerised control ever since.

6.3 Coater technology

6.3.1 Roll-coaters

6.3.1.1 Roll-coater operation

Roll-coating is the most common way of application of the liquid paint films (cf. Chapter 1.1). Obviously, the physical features of the coating material, in particular its solids content and its viscosity, and the line speed set the frame for the process. Therefore, as described previously, the applied thickness of any given coating largely depends on the settings of a roll-coater's nip pressures and roll speeds relative to the speed of the strip surface. Two- and three-roll coaters can be used in different coater designs, roll arrays and operation modes.

The most common installation of a coil coating line has the strip travelling horizontally through the processing stages. Usually, the roll-coaters therefore are mounted so that the topside coater, i.e. the one for the topcoat that is commonly awarded highest care for utmost surface quality, is operated pressing the strip against a support or back-up roll, while the reverse coater works against the floating strip from underneath. However, also arrangements of the coaters working against the strip that passes between them are known, mostly in situations with the strip travelling vertically. To achieve good coating quality in this latter case, particularly precise adjustment of the pressure of the applicator rolls against the strip is required.

Two rolls are sufficient for coating materials with lower viscosity, especially when applying thin layers. Three rolls generally enhance the accuracy of wet film control, and even more so with viscous media. As mentioned earlier (Chapter 1.1) three rolls are either arranged in a line or V-shape.

Forward operation has the applicator roll surface moving in the same direction as the coil. The speed is adjusted slightly higher than the coil's, to ensure sufficient quantities of liquid are transferred in order to maintain an undisturbed wet film. This mode is preferred because it involves a lower wear of the roll surface. On the other hand, however, it offers only a limited operational window in terms of wet film thickness.

Reverse mode application must be chosen with higher wet films, and with low-viscosity liquids. It causes substantially higher wear of the roller surface because of the cutting effect of the coil edges on the elastomer or rubber-clad roll surface.

The physical properties of the paint, like viscosity, thermal capacity and conductivity, must be adopted by appropriate coater settings, observing the temperature of the operation, as well. Also the positioning of the roll-coater in horizontal or vertical alignment must be considered, as, in a vertical arrangement, paint must not be allowed to drip down the strip.

To enable a largely undisturbed operation without stoppages or only short downtimes during coater maintenance and colour changes, a variety of solutions like double coater arrangements, transverse applicator units, shuttle coaters, and fast coupling change devices have been invented. For easier and safer exchange of rollers, automatic decouplers allow disconnecting the roll from its drive without manual intervention (Figure 6.9).

6.3.1.2 Coater rolls

Coater rolls are usually made of stainless steel, to ensure long service lives. In particular when corrosive materials are to be applied, high-grade steels are necessary. Often, the pick-up and metering rolls are equipped with additional hard chromium plating. Also ceramic rolls have been used occasionally.

Applicator rolls are clad with a rubber or elastomer material. This has to be chosen to accommodate the chemical features of the liquid, i.e. it must be solvent or water resistant, non-swelling, acid-proof, etc., according to the specific paint, lacquer or even pretreatment material that is processed. The hardness of the elastomer or rubber is chosen according to the desired wet film thickness. It usually ranges between around 45 and 60, in some applications up to 80° Shore A. Thicker films can be obtained with softer roll surfaces, while higher hardness is required when a high nip pressure must be exerted on viscous material and for low wet film gauges.

Figure 6.9: Automatic roll decoupling device. The drive spindle has been disengaged from the roll shaft Source: Globus

Mechanical wear resulting in grooved roller surfaces, due to the coil cutting into the elastomer, requires corrective measures. Whenever possible, the production of precoated coil is done in cycles from wider to smaller coils, in order to extend the useful service time of a roll. Regular inspection and, eventually, lathe machining of the roll surfaces is necessary to maintain faultless coating quality without longitudinal stripes occurring. Machining can be done several times before the elastomer cladding has to be replaced. The reduction in diameter and the resulting circumferential speed must be compensated for in the coater settings.

To obtain a superior quality of the applicator roll surface, the elastomer cladding preferably is produced by extrusion directly onto the roll core, rather than wrapping a sheet of the elastomeric material around and vulcanising it. Such sheet cladding will, no matter how accurate the final machining of the surface is done to achieve a perfect cylindrical shape, display a seam whose elastic properties are slightly different from its vicinity's. A marking of the coating film, in regular distances representing the roll circumference, and perpendicular to the strip is the result. An extruded cladding avoids this effect.

6.3.1.3 Control of the roll-coating process

Proper control of the roll-coater operation is necessary to ensure a high uniformity of the coating. In the past, the setting of a roll-coater was often done manually, making the coating process a matter of handicraft. Still good results were obtained by workers skilled in the art, helping coil coating to become the acknowledged quality technology it is today.

However, the development of higher standards and an ever better precision never stops. With the gauges of metal becoming gradually smaller, and also the growing need of applying thin films of fluids that might even have lower viscosities (e.g. pretreatments, thin-film coatings and aqueous paints), the necessity of closer control and automation became pressing.

Therefore more recently, the control of the coater rolls has received detailed attention. In modern lines, the rolls are adjusted automatically. Initial values are pre-set from databases. Roll speeds are precisely controlled to ensure harmonised rotation regardless of the actual diameter of the rolls. The coating thickness (e.g. measured by IR reflectance or X-Ray backscattering), the roll load and strip tension signals can actually be used as a loop feed for the automated coater adjustment. Sufficient accuracy of the pressure measurement is obtained by bearing or spindle measuring devices. For fine adjustment, the roll seats, mounted on high-precision linear guide ways, are positioned with step motors that allow step sizes of tenths of micrometres.

Thus, not only a very tight control of the applied film is rendered possible, but also its automated, continuous adjustment across and along the strip. The precision of this regulation is high enough to enable the liquid film application against the strip tension, rather than

Figure 6.10: Precision measuring and controlling of the coater roll setting; a) bearing load cell; b) spindle load device; c) step motors

Source: BWG

against a support (back-up) roll. It also enables high, uniform quality in vertical line designs.

Consequently, also revolving back-up rolls have been invented, that can be removed from the strip surface when appropriate during operation, without jeopardising the coating quality [13].

It has also been proposed to combine a three-roll system, its rolls mounted in a line, with an adjustable mechanism that actuates the applicator roll skewing versus the centre transfer roll. Similar equipment has been in operation in paper printing for long years. It was shown that with the controllable skew, the distribution of paint across the strip can be optimised. The skew control is performed with automated pressure control and step servo motors, hence obtaining uniform wet film thickness with deviations of less than 0.25 % [14].

Figure 6.11: Roll-coater with revolving back-up roll in a modern aluminium coil coating line Source: BWG

6.3.1.4 Auxiliary equipment

Trimmers are used to cut off small strips from the coil edges thereby ensuring precisely parallel edges and removal of minor faults and dents. The trimmer shears are guided with zero clearance, in order to keep minimum burring. By a suitable design, they also self-adjust after maintenance (grinding) without extra calibration. Swivelling trimmers allow width variations between coils under continuous operation.

Coaters and other gear that comes in direct contact with the strip need to be precisely controlled. This applies to the continuous control of the strip travel, or adjustments of the

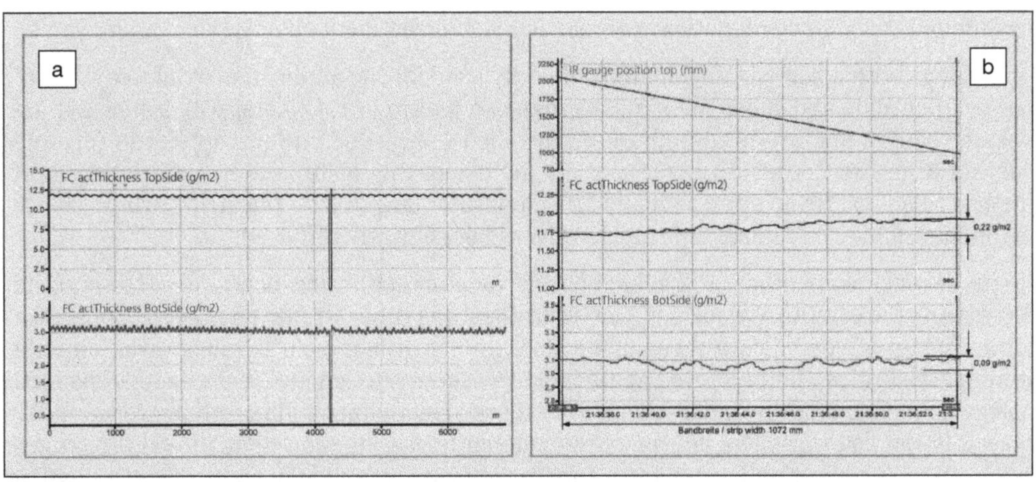

Figure 6.12: Finish coat thickness monitoring in an operation without a back-up roll;
a) aqueous top lacquer (top) and backing coat (bottom) thickness measured along the strip;
b) aqueous top lacquer (centre) and backing coat (bottom) thickness measured across the strip width; the top graph represents the position of the traversing IR head with time. The coating thickness is maintained within tenths of micrometres Source: BWG

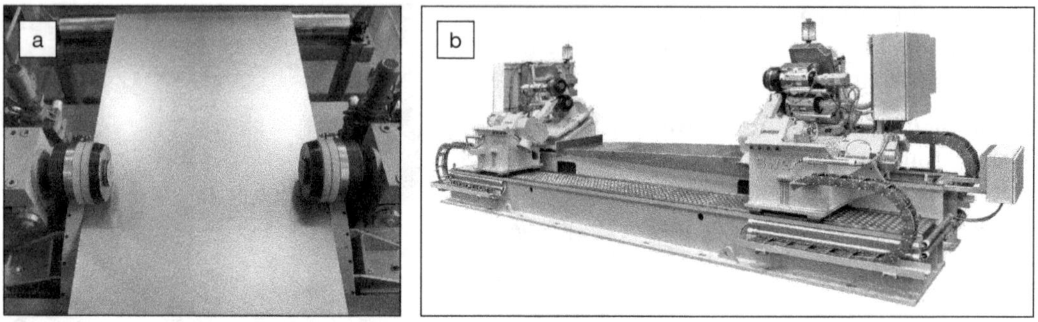

Figure 6.13: a) Side trimmer with self-recalibration in operation in an aluminium coil coating line; b) a swivelling edge trimmer allows adjusting and trimming to different coil widths without stoppage

Source: a) SMS Siemag; b) BWG

operating positions, as well as lifting off the equipment from the strip upon the seam passage between two coils. To this end, electrical, hydraulic and pneumatic systems are used as appropriate, allowing also the necessary high degree of automation.

Modern Human Machine Interfaces allow remote control and diagnosis of single equipment, but also the entire process and its safety to be done by the operators safely and comfortably from a control booth. Automatic data collection and archiving contribute to a high standard of quality control.

6.3.2 Other paint application techniques

While roll-coater technology is a long-standing and well advanced, mature technique, one of its drawbacks is the wear of roll surfaces by the contact with the moving strip and the resulting necessity for roll maintenance. Several approaches have been chosen to reduce the associated cost by using alternative application techniques.

For the application of aqueous pretreatments, instead of a roll-coater, also simpler equipment is being used, involving the spreading of a well-controlled small amount of the pretreatment liquid that is applied onto the strip a short distance before squeeze rolls. This technique is used in combination with particular pretreatment chemicals.

Applicators with a non-woven felt are commonly used for the application of lubricants (cupper lube) on strip and sheet stock in beverage can making (cf. Chapter 9.3). Issues like the passage of the chemical through the felt, controlled dosage and uniform application of aqueous wet films are not resolved yet to allow the use of this technique in coil coating. Both methods also still involve full contact of the applicator device with the strip surface. Hence, maintenance may be reduced in cost, however is not entirely removed.

A true non-contact method is provided by the vacuum applicator technique. This is transferred from the coating of wood and plastic profiles, and tubes where it has been in operation for a number of years. With these applicators, the treatment solution is sprayed onto the work piece surface, then removed again by low vacuum suction and recirculated. The technology is intended for the application of waterborne treatments and paints, as well as 100 % reactable UV coatings. The equipment is automatically adjusted to strip width changes. Non-contact (and therefore wear-free) application, wrap-around coverage and the possible high process speeds are claimed as special benefits. Pilot and commercial installations are depicted in Figure 6.14. For example, an aqueous chromium-free passivation chemical is being applied successfully since several years on a small-width hot-dip galvanised coil line. The strip width is up to 380 mm, the speed of the line, depending on the gauge of the

Table 6.1: Small capacity powder coil coating lines in Europe[18]

Company	Location	Substrate(s)	Started	Max. width [mm]	Curing	Remarks
Otefal	Bazzano (AQ), IT St.-Etienne, FR	Strip: Al/(St) Strip: Al/St	1990 1990	1500 1500	IR IR	paint/ powder powder
R. Wiegmann Metallcolor	Bersenbrück, DE	Strip/sheet: St/Al	1996	1540	IR + convection	powder
Alucoil	Miranda de Ebro (Burgos), ES	Strip: Al Strip: Al	2002 2008	1650 1650	IR IR	powder powder
Novelis	Bresso (MI), IT	Strip/sheet?: Al	1998	1720	IR	powder
Apliband	Fuenlabrada (Madrid), ES	Strip: Al Strip: Al	1998 2008	1500 1500	IR IR	powder powder
Joris IDE	Zwevezeele (W.Vlaand.), BE	Strip: St	2002	1500	NIR	powder
PIK Technologies	Moscow, RU	Strip: Al	2003	1500	IR	powder
INSI	Samara, RU	Strip: Al	?	1250	IR	powder
Elster	Lotte, DE	Strip: St	2006	500	Induction	powder
Undisclosed	Undisclosed, DE	Strip: St	2007	100	NIR	powder/EG coupled
EMB Technology	Zwolle, NL	Strip: Al/St		300	NIR	powder/pilot
Arcelor Mittal (Ligne expérimentelle des revêtements organiques)	Liège, BE	Strip: St	2003	≈ 400	diff. types	paint (roll-coater)/ powder (cloud chamber)

Figure 6.14: Vacuum applicators in operation on metal strip; a) pilot line; b) industrial coil coating line

Source: Schiele Maschinenbau

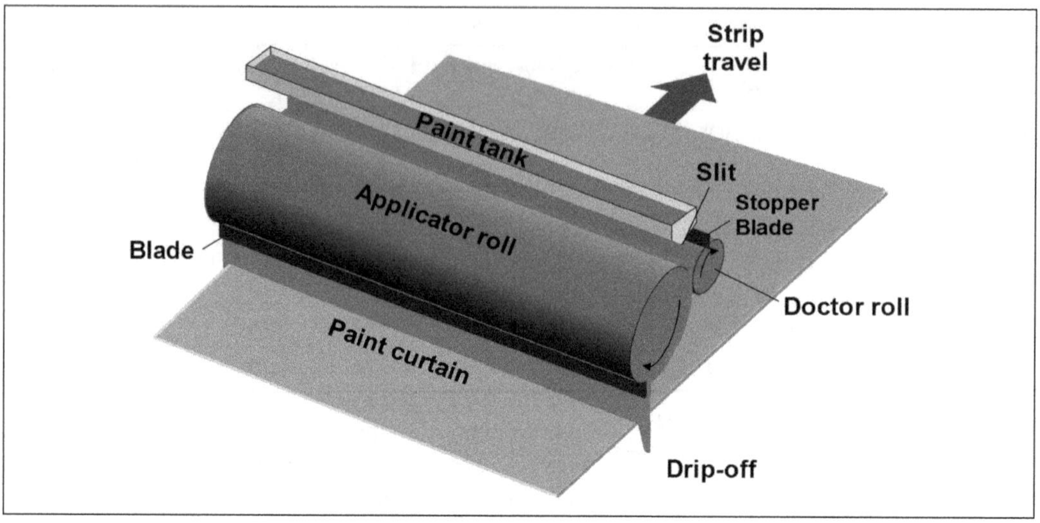

Figure 6.15: Schematic representation of a roller curtain coater

substrate, is varied between 30 and 130 m/min. The application method allows velocities up to 600 m/min [15, 16].

Techniques using the free flow of liquid paints have been studied for their suitability in coil coating application. Devices known as chambered doctor blades comprise a working tank with a bottom slit from where the paint flows, under the force of gravity, towards the workpiece surface where it is finally deposited. The thickness and uniformity of the transferred wet film is affected by the amount of paint transported through the slit - which in turn is ruled by the slit aperture, and the viscosity and temperature of the paint -, the gap between the slit and the coated surface, and the relative speed between these. The technique is industrially used e.g. in wood coating (furniture boards, doors, parquetry), or in continuous coating of paper, textiles or non-wovens. For coil coating, the principle has been adapted by removing the applicator slit a small distance from the metal surface, so that the paint passing through the slit forms a liquid curtain. Consequently, such devices are called curtain coaters.

To reliably control the amount of paint that is transferred, in a recent development, the simple tank slit has been combined with a pair of rolls that are adjustable in rotation speed and roll gap. The principle is illustrated in Figure 6.15. From the slit, the paint is taken downwards by an applicator roll, from where it is striped off by a blade, and directed to flow down to the passing strip. Paint dripping off the strip edge is collected and recirculated. The smaller diameter, counter-rotating doctor roll is used to form a stable paint curtain and to control its thickness by adjustment of the distance (gap) from the applicator roll and the differential speed of both rolls. A stopper blade mounted on the doctor roll prevents the paint from escaping from the roll nip.

The thickness and uniformity of the wet film that is formed on the strip again depends on the viscosity of the paint, the amount conveyed and the speed of the strip. For a given appliance paint system, the operational parameters of the rolls have been studied, and good consistency of the resulting dry film thickness along and across a 1.3 m wide strip is reported. The lowest film thickness that can be applied relative to the strip speed follows a hyperbolic function, hence with increased speed the resulting dry film is lower, and vice versa. Speeds between 30 and 70 m/min, and dry film gauges between 30 and 10 μm have been realised. The reported industrial line also employs induction curing (cf. Chapter 6.4.1.3) of the steel strip. Thus it was possible to encase the coater and oven units in a clean-room compartment, to securely prevent dust contamination of the delicate wet surfaces before they are cured [17].

With suitable blending, heating, and dispensing equipment, the principle of transferring a curtain of the coating material to the strip can be extended to molten paint and coatings (hotmelt), or plastic extrusion blends. Both techniques are common in other industrial uses, like adhesive bonding of paper and foil web for packaging, or the production of polyethylene and polypropylene foils. While blending, heating and transport of sufficient volumes of molten material in a line is directly transferable from other high-capacity applications, the generation of a uniform thin film on the fast moving metal strip is another matter. Reliable processing solutions, e.g. for slit nozzle arrangements in the required strip width dimensions, or the controlled formation of a film with uniform thickness are still being sought.

In recent years, a lot of attention has been paid to powder paints as they allow solvent-free operation and are acknowledged as a high-quality coating option in the metal industry. Since 1990, 13 commercial powder coating lines for coil and sheet have been put on stream in Europe (cf. Table 6.1). Though obviously being a successful business model, they are restricted to small-capacity operation due to the low process speed obtainable with current paint and application technology. Line speeds are between 6 and 30 m/min to apply the usual dry film thickness of 40 to 80 μm, and cure with combined IR/hot air ovens (cf. Chapter 6.4.1.2) [18].

Conventionally, the powder paint is applied by electrostatic spray guns onto a grounded metal workpiece. For a fast, wide web process as coil coating, this method soon meets limits in terms of transfer speed, number of spray guns and volume of pressurised air required, and overspray losses. Last but not least, the uniformity of application is critical as the powder spray from individual guns tends to create areas with lower and higher paint build. To enhance the speed and simultaneously obtain a superior uniformity of the application, several alternative technologies have been offered.

In the 1990s, an electrostatic application process for powder paints was invented that involved the powder to be applied from a wide-web feeder into a closed chamber, and directed uniformly to the metal strip by a carrier air stream through an array of deflector walls. The technology was termed PowderCloud. It allows the coating to be applied on both sides of the strip, simultaneously enhancing the speed and the uniformity of the application. The invention has been patented and licensed worldwide, however also has been the matter of a lawsuit [19, 20].

One other option is the fluidised bed method which is illustrated in Figure 6.16a and 6.16b. The powder is fed into a large tray spanning the coil width, and fluidised with air that is injected through a porous metal plate or mesh from the bottom. Applying an electrostatic high-voltage field between this mesh and a coil or sheet that passes overhead, the powder particles are accelerated towards the moving substrate surface where they are deposited in a dense layer. The thickness of this layer is highly uniform, which even can be enhanced at the metal edges (including the leading edge in case of sheet substrates) when a pulsed rather than a constant DC voltage is applied. Application speeds of up to 3 m/s (180 m/min) are reported, with marginal losses of paint through overspray, and a substantially reduced demand of process air [21, 22].

Another recently introduced powder application method makes use of the principles employed in commercial LASER printers. In this embodiment, powder paint is mixed with surface-coated magnetisable carrier particles to which the paint particles electrostatically adhere. A wide variety of suitable magnetisable, e.g. mixed iron oxide, materials are commercial.

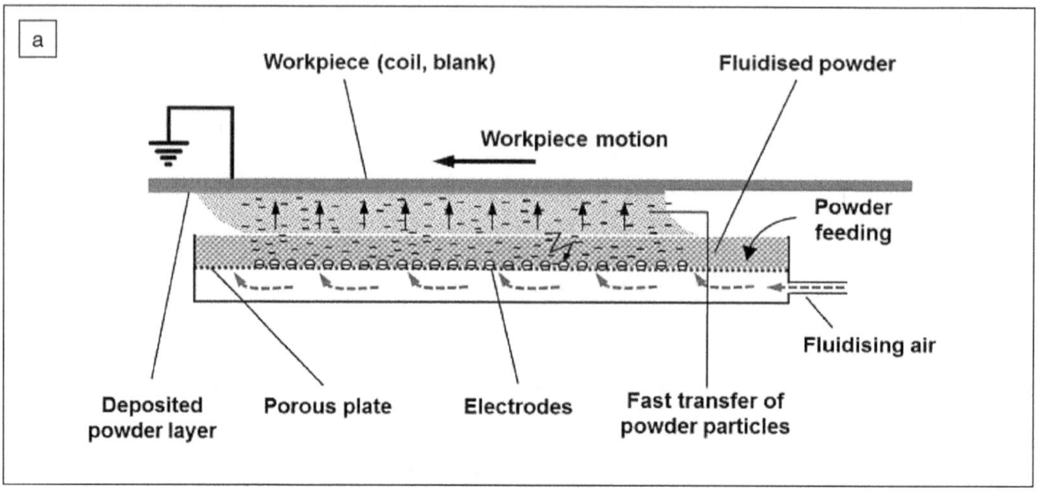

Figure 6.16: Electrostatic application of powder paint from a fluidised bed; a) principle schematic; b) pilot installation Source: Fraunhofer IPA

Examples are represented by formulas as $(Ca,Mg)Li_{0.5}Fe_{2.5}O_4$ or $CuZnFe_2O_4$. When a rotating magnetised drum is dipped into a tray containing the powder mix, it causes the magnetised oxide particles to attach and temporarily arrange into brush hairs standing out along the lines of the magnetic field. This electromagnetic brush is brought into contact with a rubber-clad roll that stripes off the powder particles and transfers them to the surface of the metal coil. The coating device can be designed for top-down and overhead coating. Currently, it is available for strip widths up to 1,600 mm. The method is reported to provide a high accuracy (± 2 %) over a wide range of coating gauges between 20 and 120 μm with virtually no powder losses. The method does not require air to carry the powder. Pilot operation is going on to establish available process speeds, working conditions and consumption data [23-25].

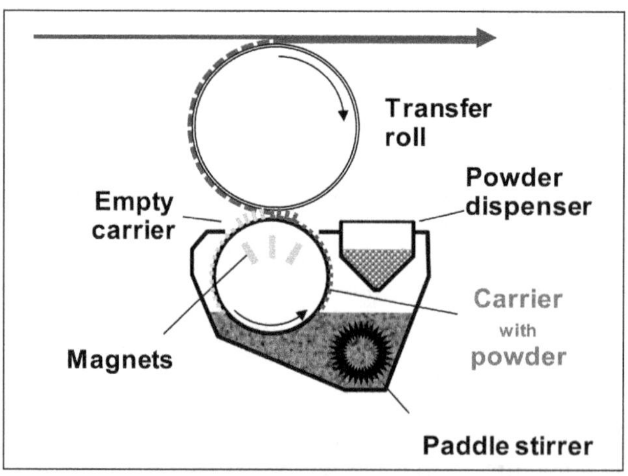

Figure 6.17: Principle of the Electro Magnetic Brush application of powder paint Illustration according to Kloppers [25]

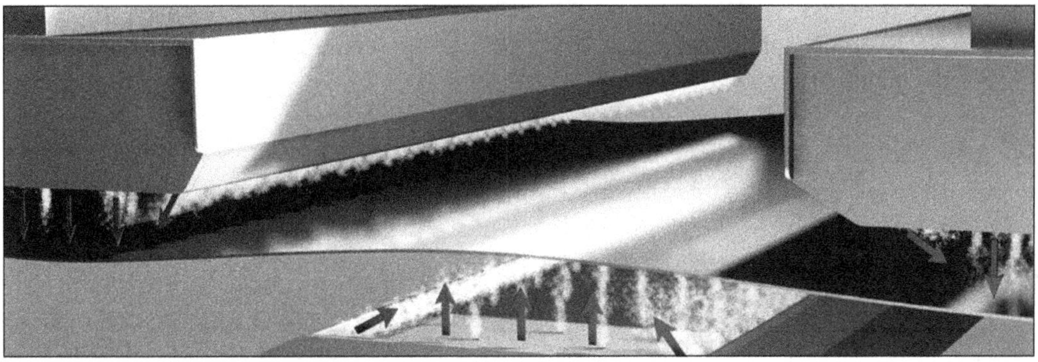

Figure 6.18: Principle of a float drier Source: BWG

6.4 Curing and oven technologies

6.4.1 Thermal curing

6.4.1.1 Convection ovens

Most coil coating lines are operated with thermal convection ovens for drying and curing. To avoid the risk of contamination by fumes, these ovens are fired indirectly (hot air, mostly by heat recovery). The design and sophistication of modern high-performance ovens ensure the automatic control and adjustment of curing parameters, so that most uniform heating and cooling of the strip are achieved, and also solvent or, in the case of aqueous coatings, water vapours are removed reliably, precluding the formation of condensation traps.

Ovens for thicker and heavier substrates are usually constructed as catenary ovens. In these, the strip hangs freely, forming a flat, inverted parabolic arch, its curvature depending on speed and tension. Blowers and nozzles inside the furnace must be adjusted accordingly, to ensure that the strip does not collide with them and a sufficient, uniform flow of hot air is maintained all over the strip surface.

Lines dedicated to thinner and light-weight substrates (e.g. for food and canstock) are often equipped with floatation ovens. These have the strip hovering on a cushion of hot air which is controlled by an array of nozzles on both sides of the strip. Uniform distribution of the oven temperature and contact-free operation are thus guaranteed. Figure 6.18 depicts the principle of floatation oven technology. The front side and control panel of a modern floatation oven are shown in Figure 6.19.

Figure 6.19: Modern floatation oven with control panel Source: SMS Siemag

Since 2008, the average utilisation of resources in coil coating lines is monitored by the European Coil Coating Association in a Sustainability Report that up to now has been issued three times. Energy usage and emissions associated to the drying, curing and incineration processes play an important role in this assessment. Table 6.2 displays the recent figures published in the 2012 Report [26].

6.4.1.2 Infrared and near-infrared curing

Table 6.2: Average resources consumption and emission data in European coil coating lines 2011 expressed as specific values per coated metal surface Data according to Alanen [26]

Resources management in coil coating lines	
Energy	
electricity	0.32 kWh/m²
fossil fuel	1.67 kWh/m²
Water	
total	2.89 l/m²
spray processes	12.70 l/m²
no rinse processes	1.70 l/m²
Solvents	
total	46.90 g/m²
incineration	39.60 g/m²
recovery	5.20 g/m²
emissions	1.50 g/m²
Scrap and waste	
scrap rate*	115.60 g/m²
waste	23.70 g/m²
recycling	15.40 g/m²
CO_2 emissions	
fossil fuel	0.587 kg/m²
electricity	0.149 kg/m²
incineration	0.091 kg/m²

* Approximated for the production of precoated steel strip (212 m² per tonne of 0.6 mm average gauge steel)

Apart from hot air convection, thermal curing of paints can be achieved by other means. Figure 6.20 illustrates how these alternative technologies use the applied energy for curing.

Infrared (IR) radiation is the portion of the electromagnetic spectrum adjacent to the lower energy, red end of the visible light band. For a representation of the electromagnetic spectrum see Chapter 5.3.3., Figure 5.1. This radiation is sensed as warmth, and in fact, is transformed, when absorbed by matter, into thermal motion, i.e. molecular translational motion, vibrations, oscillations and torsions. Thermal energy transformations are also involved in chemical reactions, as any reaction consumes or liberates energy, mostly heat, which is stored as bond energy in the chemical compounds. Furthermore, in general, molecules need

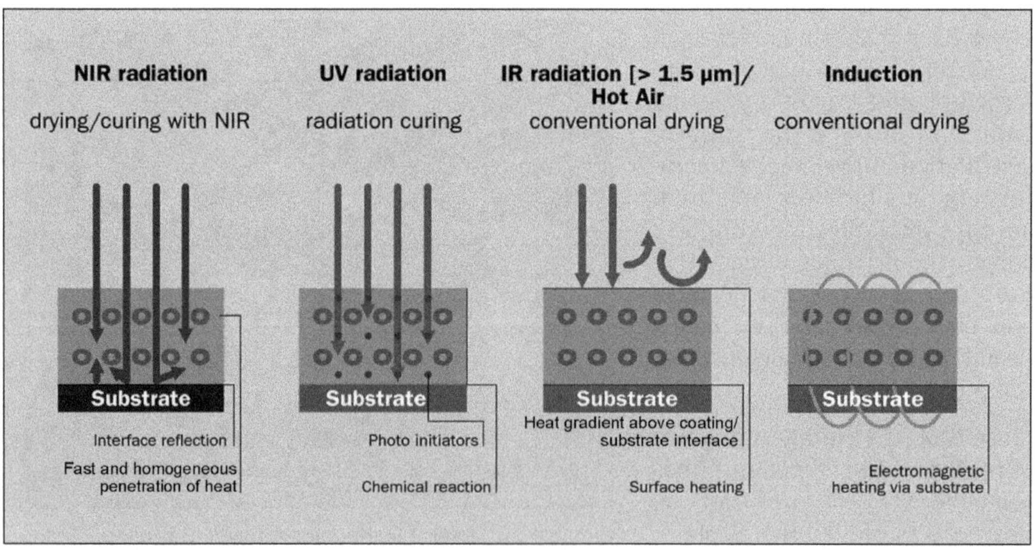

Figure 6.20: Principles of different curing technologies Source: adphos

Table 6.3: Curing data of NIR vs. conventional hot air curing of corrosion protective primers on electro-galvanised sheet

Source: adphos

	Conventional/convection Process 1. generation	NIR curing Process 1. generation	Further development	Low-cure process
Coating gauge	2.5 to 4.5 µm	2.5 to 4.5 µm	6 to 8 µm	6 to 8 µm
PMT	270°C	270 °C	260 °C	150 °C
Curing time	≈ 20 s	≈ 1.5 s	≈ 1.5 s	≈ 0.9 s
Theoret. length of installation	≈ 40 m	≈ 3.0 m	≈ 3.0 m	≈ 2.0 m

to be brought to an activated state, before any reaction, i.e. the cleavage and formation of chemical bonds, can take place. IR radiation provides suitable energy for these processes.

In industrial applications, it is particularly used for drying of aqueous systems, and the gelation and curing of powder paints. IR ovens are adjustable to individual requirements of a painting process, e.g. reflecting different paint compositions, substrates, or workpiece geometries. Their efficiency of curing is usually higher than that of hot air convection [27].

IR ovens are available with various electric, carbon, or gas-fired, ceramic radiators. These radiators are chosen according to their wavelength, their power output and emitted energy density. For example, the evaporation of water is achieved best with medium-wavelength radiation (3 µm), as this matches the absorption range of water molecules. Hence, medium-wave carbon radiators are suggested for this purpose. On the other hand, also penetration and reflection characteristics of the bulk film and the substrate have to be considered. Most wavelengths of the IR range are reflected from the surface of the liquid film or absorbed a short distance into it. As a result, a heat gradient is observed, with temperatures decreasing towards the metal/coating interface. As with convection curing, this bears a risk of prematurely curing the top of the film before solvents have entirely evaporated.

For a fast process like coil coating, preferably short-wave (approx. 2 µm) radiators are used. Carbon radiators are available also for this range. They provide transition from off or standby modes to operation within about 15 seconds, and are available with a power output that suits the purpose. Conventional medium-wave (3 to 3.5 µm) radiators are less effective, as they have a higher inertia. When comparing radiator types, it is necessary to consider the available power output, as the desired short curing time is not achieved with low energy densities [28]. On the other hand, while a certain temperature (peak metal temperature, PMT) is often used as a descriptive value for the curing process, it is not the only factor, and judgments of the curing efficiency based only on the PMT may be misleading. Also absorption, reflection and distribution of the energy, as well as the kinetics of the curing must be considered, the latter involving evaporation, deblocking of the crosslinker and polymerisation and crosslinking itself.

Near-infrared (NIR), i.e. high-energy IR radiation, uses a small band of radiation < 1.5 µm wavelength that is generated by special tungsten halogen emitters. These radiators can be switched on and off virtually instantaneously (1 to 2 s), hence are particularly suited for compact coil and sheet coating solutions (cf. Chapter 6.5). The radiation temperature corresponding to the wavelength is over 3,000 K, and specific radiation output levels of over 1 MW/m^2 are achieved. The typical specific consumption of electric energy by an NIR oven in a steel coil line is reported to be in the order of 8 to 12 kWh per tonne of steel treated.

NIR provides the energy that matches the evaporation and reaction enthalpies of solvent or waterborne paints. It penetrates the liquid coating homogeneously, and is reflected at the surface of the substrate, allowing the removal of solvents or water from paints, as well as cross-linking reactions to complete within a few seconds. Thus, curing can be achieved substantially faster than with a convection oven, accordingly resulting in a much shorter oven length (cf. Table 6.3). In combination with a suitable airflow management, this curing speed is unchallenged, with the exception of UV curing [29].

In general, polyester/melamine coil coating paints are cured well and with a good surface quality in well under 10 s. Yet it has been shown that, as pigments absorb the NIR radiation differently, adjustment of the irradiation intensity becomes necessary, hence power and effective dwell time of the NIR device must be manipulated. Especially dark pigmentation (carbon black) absorbs NIR well resulting in overdue heating which threatens to bring on crosslinking before the solvents have entirely been evaporated. Hence, there is a risk of boiling blisters. Similarly, metallic paints containing aluminium platelets, despite a high reflectance, are heated up to suffer premature crosslinking. It has been argued that these absorption and reflection effects form an inherent problem of NIR curing restricting its applicability [30]. However, one way to cope with these effects is the stepwise, intensity controlled irradiation and an interval between the NIR heads. The temperature obtained by the coil serves as feedback. Control of the NIR transparency of the wet coating film by suitable choice of pigments and their volume concentration is another means of dealing with the situation. For example, non-platelet Al flakes may even enhance the curing process, as they cause high reflection, hence effective energy distribution, inside the curing film.

NIR curing has proliferated in coil coating since the year 2000. Many recently built high-capacity coil coating lines and combined continuous galvanising/coating lines all over the world have been equipped with NIR furnaces. Short-built NIR modules are also used to boost the capacity of existing convection ovens. For the curing of conventional powder paints in coil and sheet installations, line speeds of 60 m/min and very short curing times can be achieved. In particular, powders based on a novel uretdione approach, have been cured with curing times below 5 s and particularly low energy consumption below 0.4 kWh/m^2 [31].

6.4.1.3 Induction curing

Drying and curing is also possible by applying a high-frequency, e.g. 5 to 50 kHz, alternating current (AC) to inductor coils wound around the metal strip. The AC spontaneously induces eddy currents in the metal, generating resistive heat in turn. Heating rates that are sufficient for coil coating require magnetisable substrates, particularly carbon steel, where magnetic hysteresis losses add to the effect. Consequently, induction ovens have been installed in operations that exclusively process steel substrates. Dedicated coil coating, passivation and coating sections in galvanising lines, as well as varnish coating lines for silicon steel are exemplary applications. The short curing times of about 10 s achievable by the technology, and the adaptable design of the oven tunnel make for a convenient use of induction curing in vertical installations. Combinations of induction ovens (boosters) with conventional convection ovens are also reported.

To enable a uniform magnetic flux and avoid sparking, special inductor designs are common. Usually, several consecutive inductors are mounted that heat up the substrate stepwise. It is recommended to install longer individual inductor sections that can be operated at the lower end of the mentioned frequency range, i.e. 5 to 10 kHz. Thus, heating rates can be kept to a preferable level of e.g. 30 K/min, and overheating can be avoided, especially at strip edges where the magnetic field is strongest. Other than convection or (near) infrared curing, induction curing results in a reverted temperature gradient, as the heating energy stems from the substrate metal. Heating is usually supported by air re-injected from the incinerator system, to provide a sufficient air flow and control the solvent vapour concentration in the exhaust below the lower explosive limit (LEL, cf. Chapter 6.4.3). Modern inductor tunnel design helps to accommodate laminar airflow and to avoid the formation of condensation traps [32, 33].

6.4.1.4 Comparison of thermal curing techniques

With the requirement of increased process speed, alternative curing options to the well-established convection ovens are being considered. It is argued that the transfer of energy by IR radiation or inductive heating generally requires shorter dwell times for drying and curing, and heating of large volumes of air is avoided. In fact, convection curing involves a lot of waste heat that is lost by dissipation if not recuperated for other purposes. Moreover, hot-air ovens, due to their inertia, require long idle times and dummy coil operation whenever the dimensions of the coils change, to adapt to the heat capacity that changes accordingly.

IR and induction ovens can be instantly switched between standby and operation modes. As IR sources can be controlled individually, a web of emitters mounted across the strip width enables adjustment to width variations between coils.

Figure 6.21 shows heating curves obtained with different thermal curing techniques. For the illustration, typical oven layouts in coil coating lines have been investigated. NIR modules enable the steepest heating curve. As mentioned above, it is useful to apply the radiation in two steps. The first NIR head is used for evaporation of solvents. Due to its heat capacity, the coated coil will slowly take up more temperature during the interval. To enhance evaporation of solvents and/or activation of blocked cross-linking agents, the NIR heads can even be separated farther to extend the dwell time. In a patented "active dwell time" design, the coil is treated with recirculated warm air. The air flow is controlled observing the lower explosive limit (LEL, cf. Chapter 6.4.3). The second NIR head brings about the cross-linking [34, 35].

A characteristic feature of induction ovens is that heating of the metal occurs only in the electro-magnetic field of the inductors. Oven designs with multiple inductors therefore

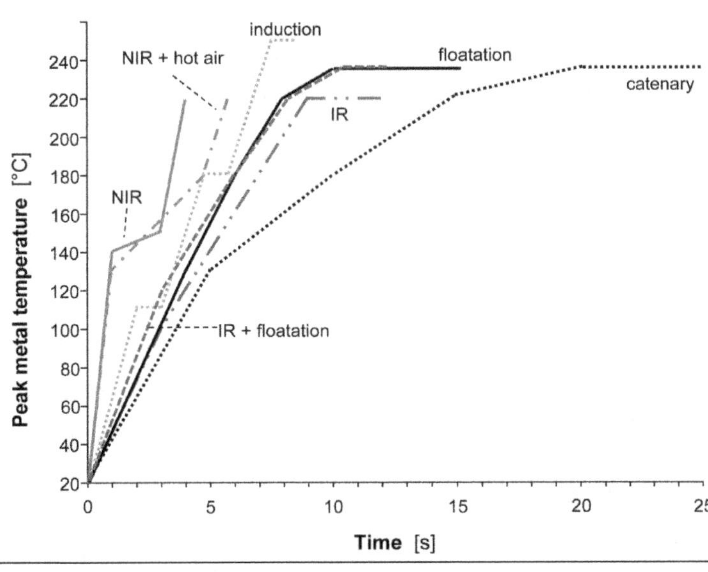

Figure 6.21: Typical heating curves of thermal curing ovens Source: Bär [29]

Table 6.4: Comparative data for thermal curing technologies in coil coating operations. Consumption figures are taken from customer audit and published data Data acc. to Alanen[36] and Bär[37]

	Convection	IR	IR + convection	NIR	Induction
Speed limit [m/min]	180	horizontal: 180 vertical: 150	horizontal: 180 vertical: 120	n.a.	horizontal: < 180 vertical: 150
Min oven dwell time [s] < 12 µm dry film 12 to 20 µm 20 to 25 µm	 12 to 15 15 to 18 18 to 20	 8 to 10 10 to 13 13 to 15	 ≥ 8 > 10 > 12	 < 3 3 to 4 4 to 6	 6 to 8 8 to 10 10 to 12
Dynamic operation (s)	several minutes	30 to 120	30 to 120	1 to 2	10
Start/standby requirements	high	low	medium	n.a.	n.a.
PMT [°C]	225 to 235	225 to 235	225 to 255	200 to 215	225 to 240
Energy source	gas & electricity	electricity	electricity, gas & clean air from RTO	electricity & optionally clean air from RTO	electricity
Energy consumption [kWh/t] Gas Electricity	 354 68	 25 200 to 250	 180 110 to 130	 < 5 130 to 150	 25 180 to 200
Exhaust treatment	any	any	any	RTO preferred	any

have a typical step-up heating curve. In convection ovens, the metal strip attains its final temperature in an asymptotic curve, depending on the air temperature.

Table 6.4 displays results of a comparison of thermal curing technologies in terms of obtainable process speeds and curing times, dynamic process characteristics, typical curing temperatures (PMT) for a standard polyester/melamine paint system, and energy and exhaust treatment requirements. Consumption data for conventional hot air convection ovens refer to the ECCA Sustainability Report of 2012, approximating the data for the production of precoated steel strip (212 m^2 per tonne of 0.6 mm gauge steel) from the given average total value [36]. Other consumption figures are reported from customer audits.

It has been argued that further potential exists for optimisation [37, 38]. For instance, the use of fossil fuel (natural gas) for the incineration is likely to be substantially reduced when employing any of the infrared radiation or induction curing techniques. As a consequence, the related emission of CO_2, either through direct incineration or as equivalent from the generation of electrical power, is largely diminished as well. Reduced installation length due to the faster curing, shorter idle times during strip dimension changes, resulting in capacity increases and less scrap from dummy coil operation, lower energy consumption and a lower CO_2 footprint all add up to potential savings in the order of several hundred thousand to a million Euro per year in an average line of 100 kt capacity.

6.4.2 Radiation curing

An introduction into UV curing is given in Chapter 5.3.3. UV curing requires special formulations based on reactable monomers and photoinitiators that cure via a photopolymerisation process. Especially the radical type of photopolymerisation reactions provides a very high curing speed. UV formulations that contain solvent or water to control their viscosity require complete evaporation of these carriers in a separate oven before the curing is done.

High-pressure UV equipment is available with high energy output as is required for the fast and wide-web operation of a coil coating line. Typical mercury (Hg) vapour lamps with an output of 2 W/cm^2 provide a sufficient energy density for high-speed curing. A range of industrial suppliers offer total systems including lamps, power supply and amplifiers, reflectors and shutter systems as well as cooling and air/nitrogen supply [39, 40]. To-date, UV curing installations have only found a few applications in organic coil coating lines, two of them specialising in transparent coatings for stainless steel.

EB involves accelerated electrons that are generated in cathode tubes at high voltage. The electron source comprises a high-vacuum chamber with an array of cathodes, control and anode grids, and a water-cooled screen with a titanium foil that, though being vacuum-tight, serves as a window for the accelerated electrons to escape. Due to their high speed of half the velocity of light, a sufficient amount of electrons passes both the anode grid and the window foil, to hit and penetrate the target. The intensity of the electron beam is controlled by the acceleration voltage and the voltage of the control grid that rules the electron current. Hence, the further equipment includes power supply to generate the radiation of up to 250 keV, and the adjustment systems for the control grid [41].

In industrial line operation, safety precautions like proper optical and X-Ray shielding are mandatory.

6.4.3 Cooling and exhaust management

As the strip obtains a temperature that may be as high as 250 °C (e.g. PVdF coatings, see Chapter 5.5.3), cooling of the strip is necessary once the curing is achieved. This is done by water quench systems, cooling rolls or air cooling, depending on the temperature difference and the heat that has to be covered. Quite usually, combinations of these techniques are used. It is also common to use the hot water from the quench for the heating of process baths etc.

Incinerator systems are required to eliminate solvent vapours from the exhaust whose temperatures are usually 250 to 350 °C. To achieve this, a temperature of > 750 °C is needed, which normally requires the addition of fuel (natural gas). Modern sophisticated systems are designed and operated to minimise the natural gas consumption, and to recuperate or regenerate the heat [42].

As chemical energy is stored in the organic solvent molecules, this energy is liberated by the burning process, so that the solvent itself forms the fuel. In this so-called autothermal operation of the incinerator, no extra fuel is needed to maintain the burning. The energy consumption of the curing process is hence optimised. Modern incinerators work on the principle of preheating the exhaust from the curing ovens by heat exchange with the even hotter incinerator exhaust. The most advanced ceramic heat exchangers are capable of achieving a temperature close to the self-ignition point, so that virtually no extra fuel is required except during start-up and idle times. Incinerators using this technique are therefore commonly referred to as Regenerative Thermal Oxidisers (RTO). Recently installed incinerators are shown in Figure 6.22a and 6.22b.

The operation of the curing oven and incinerator of course has to obey safety levels in terms of the allowed solvent content in the exhaust that is to be treated. Any flammable substance has a specific concentration in air, either as gas or aerosol, that defines its lower explosive

Figure 6.22: Modern afterburner systems; a) three-chamber regenerative combustion oven; b) recuperative system with primary and secondary heat exchanger Source: a) SMS Siemag; b) Globus

limit (LEL). Usually, a concentration of 25 % of this level is considered the safety limit, under worst-case conditions, in order to avoid uncontrolled combustion. The exhaust therefore needs to be diluted with air, which again, is preheated by heat exchange to save energy. This is done by measurement of the vapour concentration by gas sensors, and automatic feeding of the required amount of air to maintain the 25 % LEL level [43].

Overall, a modern, safe layout operates at stack temperatures reduced by the heat regeneration to 150 °C or even below. An emission level of less than 20 mg carbon per normal cubic meter, as required by environmental directives, is easily achieved [44].

Oven and incinerator operation are supported by algorithmic control programs that calculate air flow, temperature curve and distribution, and set oven parameters as a funtion of substrate characteristics and the size and speed of the strip. This computerised control obviously as well enables the creation and use of stored data for reliable first-pass operation.

6.5 Examples of modern coil coating installations

As technology development is a continuous process, the decision which features make a coil coating line a modern installation is always somewhat arbitrary. Thus the list in Table 6.5 is by no means exhaustive. It shows a very brief selection of high-capacity lines that involve most of the modern engineering and processing concepts described earlier in this chapter. Lines using this technology are also exported overseas. All of the lines mentioned have been commissioned or substantially refurbished during the last decade, or will be put on stream in the next few months [45-52].

The currently discussed concept of Compact Coil Coating is dealing with combining compact installations and processes, to enable the cost-efficient production of precoated metal in small batches. A suitable definition might be that such a line is capable of economically producing order sizes between 200 kg and 5 t, thus far below the usual minimum-order size of 20 t.

Most of the European lines that already comply with this definition for a compact line are restricted to narrow strip, e.g. for the production of stock for Venetian blinds. The novel, compact and lean concept of coil coating however targets the production of wide strip. It has been argued that such a concept would offer an excellent business opportunity for service centres and job coaters, but also for metal workers that up to now have been forced to operate a post-coating process.

The new concept embraces start-stop operation, thereby allowing quality coating of metal coil from the first metre with minimised scrap. One precondition for this concept is the

Table 6.5: Selected installations and projects of modern coil coating lines under construction; data according to [46–53]

Line	Country	Substrate	Width (mm)	Gauge (mm)	Speed (m/min)	Commis-sioned	Remarks
Alcoa, Merxheim	F	Al, Al composite	2,000	0.3 to 2.0		2008	reversing backup roll, floatation oven, RTO
Alcoa, Samara	RUS	Al	2,050	0.15 to 0.8	250	2009	swivel edge trimmer, reversing backup roll, floatation oven, RTO
ASAS Alüminyum, Akyazi	TR	St, Al	1,600 / 2,100	0.2 to 1.0 / 2.0	80[a]	2013	
Henan Zhongfu, Gongyi Cty.	CHN	Al	1,900	0.15 to 1.0	250	2014	inline stretcher-leveller, floatation oven, regenerative exhaust combustion
Marce-gaglia, Ravenna	I	galv. steel	1,650	0.25 to 1.6	180	2012	inline coating, NIR oven, drip-protection
Thyssen-Krupp Steel, Duisburg	D	electro-galv. steel	1,900	0.5 to 1.75	120[b]	2003[b]	inline coating, NIR oven
Thyssen-Krupp Steel, Kreuztal-Eichen	D	St., (electro-) galv. steel	1,750	0.5 to 3.0	120	2012[c]	refurbishment: catenary oven, RTO

[a] *Process section;* [b] *inline coating section;* [c] *oven refurbishment*

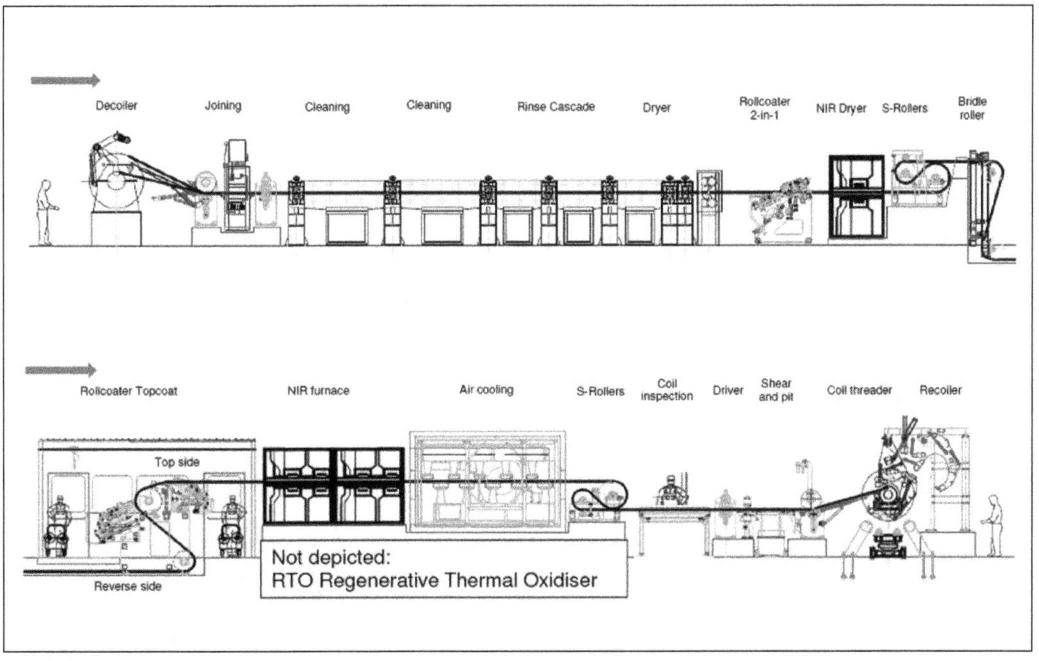

Figure 6.23: Concept design of a compact coil coating line employing 2-in-1 pretreatment primer coating, NIR curing and start-stop operation

Source: *adphos/Globus*

availability of fast-response curing technologies like near infrared. Another key factor of an industrially operable start-stop technology is a centralised control unit and software for simultaneous output monitoring and power control of the individual emitters of the curing oven. Both technologies already have been introduced in a variety of industrial lines, while the third component, the 2-in-1 primer-pretreatment technology, has entered the market in but a few places (cf. Chapter 4.5.2). Figure 6.23 illustrates a projected compact coil coating line [54].

6.6 Literature

[1] Alanen, A., ECCA Sustainability Report, ECCA, Brussels 2012

[2] anon., Precision Spray Nozzles for the Chemical Industry, Brochure, Lechler GmbH, Metzingen 2012

[3] anon., Flat fan nozzles, Lechler GmbH, Metzingen 2013, www.lechler.com/products/general industry/flat fan nozzles (11-09-13, 19:40 h)

[4] Krüger, R., Weltweit erste Coil Coating Anlage mit Inline-Plasmareinigung läuft (First coil coating line ... with inline plasma cleaning...), besserlackieren 9, 2007, pp. 1 ff

[5] Kotte, L., Mäder, G., Althues, H., Kaskel, S., Roch, J., Leupolt, B., Einsatz großflächiger Lichtbogen-Plasmatechnologie zur Erhöhung der Haftfestigkeit (Use of large-area electric arc plasma technology...), Workshop ak-adp Dortmund, Proc., MEOX Projektmanagement, Jena 2013

[6] Lebelt, P., Gehrke, J., Hallmeier, K. H., Reuther, H., Dörfel, I., Qualitative und quantitative Analysen dünner Funktionsschichten für das Kleben durch XPS-, AES-, REM- und TEM-Untersuchungen (Qualitative and Quantitative Analyses of Thin Functional Coatings...), Galvanotechnik 107, 2009, p. 930 ff

[7] Schinkinger, Schichtanalytische und elektrochemische Untersuchungen zur Abscheidung dünner SiO_2-und Organosilanschichten auf verzinktem Stahl (...Studies on the deposition of thin SiO_2 and organosilane coatings on galvanised steel), Diss., Ruhr-Univ. Bochum 2004, pp. 32 f; 65 ff; 100 ff; 140 ff

[8] Schimanski, A., Oberflächenfunktionalisierung bei Atmosphärendruck - Verfahren und Möglichkeiten, Surface functionalisation at atmospheric pressure...) Workshop ak-adp Dortmund, Proc., MEOX Projektmanagement, Jena 2013

[9] Tiller, C., private communication, Sura Instruments GmbH, Jena 2013

[10] Koppers, H., Beuther, H., Sonntag, S., Hoffmann, J., Noé, A., BWG 50 Jahre - 50 Years, BWG Duisburg 2005, pp. 278 ff

[11] Rieth, B., Successful modernisation of Russian rolling plants, Aluminium 85, 2009, pp. 34 ff

[12] anon., BWG Supplies State-of-the-Art Coil Coating Line to Russia, Brochure, BWG, Duisburg 2009, p.3

[13] Rieth, B. [11], p. 38

[14] Bridges, M., A new approach in coating equipment design to improve cross-machine paint thickness profile, ECCA Autumn Congress, Proc., European Coil Coating Association, Brussels 2009

[15] anon., Products: Vacumat: Vacumat Special Coat, Schiele Maschinenbau GmbH, Niederzissen 2011; www.schiele-maschinenbau.de/... (20-06-13; 12:14 h)

[16] Brisberger, R., private communication, Wuppermann GmbH, A-Judenburg 2013

[17] Ueda, K., Application of Roller Curtain Coater for the Steel Strip, ECCA Autumn Congress Brussels, Proc., European Coil Coating Association, Brussels 2012

[18] Bielefeld, F.W., Sander, J., The profitable choice - From small-batch post painting to compact coil coating, Europ. Coatings Conf. Market Day Coil and Can Coatings, Proc., Vincentz Network, Berlin 2011

[19] Escallon, E.C. (inv.), Electrostatic powder coating apparatus and method, US 5695826, Terronics Development Corp., Elwood (IN), USA 1997; www.espacenet.com/...; 16-09-13, 11.05 h

[20] Cudahy, U.S. Circuit Judge, SMS Demag Aktiengesellschaft v. Material Sciences Corporation, Judgment, United States Court of Appeals, Seventh Circuit, 2009; http://caselaw.findlaw.com/us-7th-circuit/1268634.html; 16-09-13, 10.04 h

[21] Cudazzo, M., Strohbeck, U., Beckmann, M., Hermann, S. (inv.), Device and method for coating of a flat piece, WO 2004091805 A1, Fraunhofer Ges., München 2004; www.espacenet.com/...; 16-09-13, 10.26 h

[22] Gedan-Smolka, M., Strohbeck, U., Demmler, M., Cudazzo, M., Vogelsang, H., Tuschla, M., Ressourcenschonendes Pulver-Precoating-Verfahren eröffnet neue Märkte für vorbeschichtetes Blech

(Resources saving powder pre-coating process…), 77. GDCh Lacktagung, Proc., Ges. Dtsch. Chem. (German Chem. Soc.), Frankfurt a.M. 2012

[23] Rickert, C., Kress, J., Probst, M. (inv.), Process for the Transfer of Powders and Powder Coatings to Substrates, US 7615256, Atotech Deutschland GmbH, Berlin 2008; www.espacenet.com/...; 16-09-13, 12.57 h

[24] Kloppers, P., A new powder application system: EMB - Electro Magnetic Brush, 8th International NIR Symposium, Proc., adphos, Bruckmühl 2011

[25] Kloppers, P., EMB Status Update, ECCA Autumn Congress, Proc., European Coil Coating Association, Brussels 2009

[26] Alanen, A. [1]

[27] anon., Infrarot-Strahler für Beschichtungen (Infrared radiators for coatings), http://www.heraeus-noblelight.com/de/applications/infrarot/coating_1/ir_coating.aspx; 25-05-13, 12.34h

[28] Gedan-Smolka, M. et al. [22]

[29] Bär, K. K. O., Instantaneous and ultra-fast curing with NIR, 8th International NIR Symposium, Proc., adphos, Bruckmühl 2011

[30] Mabbett, I., Elvins, J., Gowenlock, C., Jones, P., Worsley, D., Effects of highly absorbing pigments on near infrared cured polyester/melamine coil coatings, Progr. Org. Coatgs. 76, 2013, pp. 1184 ff

[31] Gedan-Smolka, M. et al [22]

[32] Delaunay, D., Induction heating benefits in coil coating, ECCA Autumn Congress, Proc., European Coil Coating Association, Brussels 2001

[33] Brassel, G., Lovens, J., Unicoat Gen 5 - The Inductotherm Curing System installed on the CGL7 by ArcelorMittal in Yvoz-Ramet (Belgium), ECCA Autumn Congress, Proc., European Coil Coating Association, Brussels 2012

[34] Bär, K. K. O. [29]

[35] Mabbett, I. et al. [30], pp. 1189 f

[36] Alanen, A. [1]

[37] Bär, K.K.O., Clean Technologies in Coil Coating, ECCA Autumn Congress, Proc., ECCA, Brussels 2011

[38] Bär, K.K.O., New opportunities, added value and lower cost coil coating with strong reduced emissions, ECCA Autumn Congress, Proc., European Coil Coating Association, Brussels 2012

[39] Jung, J., Härtung von Druckfarben und Lacken durch UV-Strahlung (Curing of Printing Inks and Paints by UV Radiation), IST-Metz GmbH, in: Dtsch. Berufsgenossenschaft Energie Textil Elektro Medienerzeugnisse (Institution for Statutory Accident Insurance and Prevention - Energy Textile Electro Media Products, Working Committee for UV printing) BG ETEM (ed.), UV Technology - A Practical Guide for all Printing Processes, BG ETEM, Wiesbaden 2007, pp. 29 ff

[40] Skinner, D., New Development in UV Curing Systems for the Coil Coating Industry, ECCA Autumn Congress, Proc., European Coil Coating Association, Brussels 2004

[41] Lowe, C., WeatherEbility, ECCA Autumn Congress, Proc., ECCA, Brussels 2009

[42] Banabak, H., Minimizing energy usage - A case study, ECCA Autumn Congress, Proc., European Coil Coating Association, Brussels 2010

[43] anon., Explosion protection, Brochure, Dräger Safety AG & Co.KGaA, Lübeck 2013; http://www.draeger.com/sites/assets/PublishingImages/Segments/NL/Applications/Plant-Safety-and-Operations/explosion_protection_br_9046262_en.pdf; 18-09-13, 16:14 h

[44] anon., Oven changed within four weeks, Newsletter, SMS Siemag AG, Düsseldorf 2013

[45] Rieth, B., One of the most modern coil coating lines for aluminium, Aluminium 84, 2008, pp. 40 ff

[46] Rieth, B. [11], pp. 34 ff

[47] anon. [12]

[48] Sasse, C., private communication, SMS Siemag AG, Düsseldorf 2013

[49] anon., Henan Zhongfu relies on color coating line made by SMS Siemag, Newsletter, SMS Siemag AG, Düsseldorf 2012

[50] Bär, K.K.O., Bandprozesse in einem Fluss (Coil processes in a flow), JOT (7) 2012, pp. 16 ff

[51] Koppers, H. et al. [10], p. 351 ff

[52] anon. [44]

[53] Bielefeld, F.W., Sander, J. [18]

7 Performance testing

7.1 Introduction

Testing of industrial goods serves the purpose of defining specifications and verifying their observance by quality control. General tests for coated metals are related to features like cosmetic properties (colour, gloss), hardness, resistance to abrasion, and in particular the total thickness of a coating. In addition to that a large number of tests have been invented to assess the performance of coated metal. The following chapter describes the most common procedures and the related standards for the performance testing of an organic coil coating (cf. Chapter 10.5.4 for a list of standards).

7.2 Coating thickness

Coatings on coil and sheet metal are applied in a thickness range of between approximately 3 µm (e.g. primers, anti-fingerprint coatings, clearcoats) and 200 µm or more (PVC plastisols). Commonly, the dry film thickness can be measured by applying a conical bore through the paint film and measuring the thickness under the microscope. It is also possible to prepare a cross-section and measure the gauge of superficial layers by electron micrography (SEM), though this method is too complex and costly to serve as a tool in everyday operation. Other methods employ induction or eddy current measurements [1]. A common method for the detection and quantification of pretreatments makes use of X-Ray fluorescence that is given by marker or tracer elements present in the coating. Most of these methods require a test specimen to be cut from the coated metal. However, also continuous online methods have been invented. An automated method for dry film measurement has been described that is based on the creation of tiny holes, just below visibility, by LASER ablation, and measurement of their depth. According to the author, the holes do not affect the performance of the coated coil. In the same paper, a non-destructive method for wet film measurement was proposed that operates with confocal optical sensors [2].

For non-destructive measurement of dry film thickness, heat reflection and β-back-scattering measurements are in use. Both have an extremely fast response, and hence can be conveniently applied for real-time operation control in coil coating lines. The heat reflection method involves local heating of the coating by a pulsed LASER beam. The heat reflected from the metal surface experiences a phase shift corresponding to the coating thickness. Measurement of this reflected heat wave with an IR photo sensor and algorithmic processing of the data renders a highly accurate (± 0.1 µm) result [3].

Gauge measurement through the detection of backscattered β-particles is a long-standing method. Slow electrons emitted by a β-radioactive source like the ^{85}Kr isotope are reflected (backscattered) from the coated coil surface, and measured with a Geiger counter. The rate of backscattering is representative of the coating thickness [4]. IR reflectance methods have been used in the past mainly for wet film measurement, using the strong absorption of O-H and C-H bonds around 3,000 cm^{-1}. Improved accuracy of the detectors and elaborate data evaluation and processing meanwhile also enable reliable dry film monitoring [5].

7.3 Adhesion testing

7.3.1 Overview

Adhesion of an organic coating, as has been discussed already, is absolutely essential. This is particularly true for precoated metal, as this use implies severe deformation during the processing and transformation into 3-D final goods. Adhesion tests can be carried out by direct measurement, e.g. by pull-off or torsion shearing. More common tests for coil coated sheet involve the simulation of deformations encountered during metalworking and final use. Table 7.1 provides a brief description of such deformation tests in coil coating. Table 7.2 gives the relevant European standards for these and selected further tests, which generally have been pooled in the EN 13523 standard series (cf. Chapter 10.5.4.3).

7.3.2 Application tests

7.3.2.1 Tests involving deformation

For coil coated metal, the adhesion of a paint film under deformation is the most critical mechanical requirement to achieve. Industrial tests related to the adherence of paint films

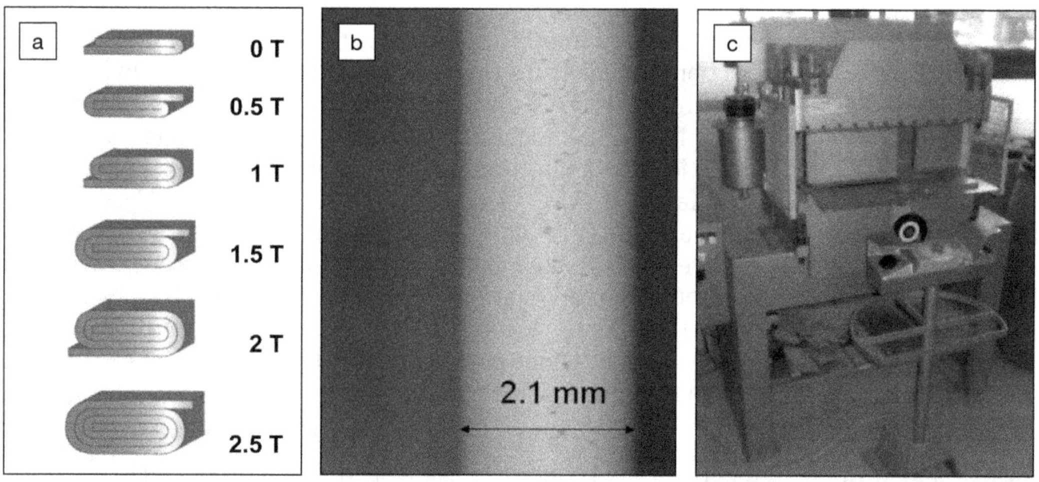

Figure 7.1: a) Annotation of T-bend test results according to the European standard, EN 13523-7; b) microscopic view of a taped T-bend shoulder, 0.5T, with paint cracks; c) hydraulic T-bending machine

Source: a) Arcelor Mittal; b) Henkel

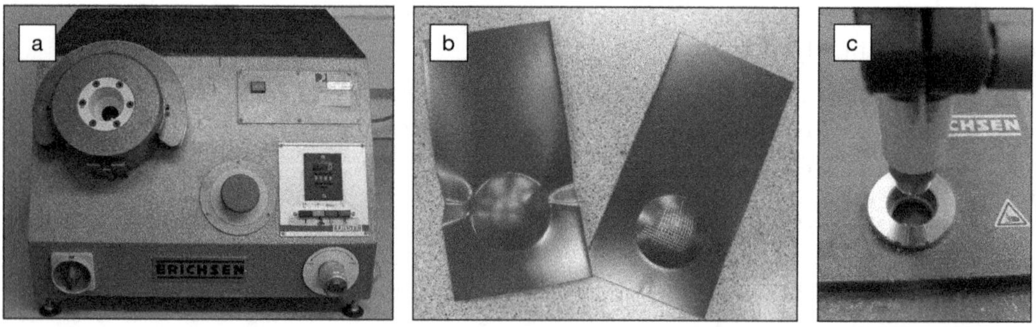

Figure 7.2: Mechanical testing; a) hydraulic Erichsen indenter machine; a stereomicroscope can be mounted for inspection; b) coated test panels with18 J reverse impact (left) and 8 mm Erichsen dome/1 mm cross-hatch applied; c) ball impact indenter tip

therefore employ standardised, slow or fast deformation events [6].

Table 7.1: Deformation tests to simulate mechanical stress in coil coating

Mandrel bending, T-bend tests: Bending around cylindrical or conical mandrels, or panel edge, assessing the minimum bending radius before failure
Erichsen cupping test: Slow drawing of a spherical cup, assessment of dome height before failure, often combined with cross-hatch testing
Other drawing tests: Drawing of cylindrical/quadrangular cups, food or beverage cans, assessment of drawing force and wear
Ball impact test: Impact of a weight with small spherical tip, measurement of impact energy before failure (top and reverse)
Bend-and-impact test: Impact of a weight with tapering edge, measurement of the distance between metal rupture and paint delamination

The most common test in coil coating is T-bending, i.e. the simulation of flanging of metal sheet. In this test, the sharpest bend is denoted that is possible before the paint cracks (elasticity) or even delaminates (adhesion) from the bent shoulder. The inner radius achieved is annotated in units of the sheet thickness. A bending value of 0T denotes the panel bent back on itself; bending around the edge of another sheet of the same thickness is annotated as a value of 0.5T; and so on (EN 13523-7; cf. Figure 7.1a and b). Since the corresponding American standard (ASTM D-4145-83) counts the inserted sheets, which results in double T-values, reference must be made to the actually used standard when results are communicated. Figure 7.1c shows a hydraulic sheet metal folder with adjusted dies to obtain reproducible T-bends.

Table 7.2: European standards for adhesion tests of particular importance for coil coating Data acc. to Meuthen, Jandel [6]

Adhesion test standards	
• Cross-hatch test	EN ISO 2409
• Peel Tests, Pull-off Test	EN ISO 4624
• Impact; cracking resistance to fast deformation	EN 13523-5
• Erichsen dome test (cupping, slow deformation)	EN 13523-6
• T-Bending; Coating elasticity, crack resistance; Substrate adhesion, adhesive tape	EN 13523-7
• MEK Test; curing (cross-linking) and adhesion	EN 13523-11
• Heat resistance; mechanical effect of ageing	EN 13523-13
• Foil adhesion, protective foils	EN 13523-17

Slow deformation by the Erichsen dome test (Erichsen indentation, cf. Figure 7.2) is used to assess the deep-drawability of coil coated sheet. A punch with a 20 mm spherical tip is pressed into the sheet, which the standard requires to be done at uniform speed. The paint delamination and cracking are assessed from either flanks or top of the dome that forms. The test is often combined with a cross-hatch, and inspection made after taping. Usually, the test is performed to an agreed dome height, considering the formability of the base metal, to avoid rupture of the metal itself (EN 13523-6).

Deep-drawing tests are also performed by drawing of cups of cylindrical or quadrangular shape, model cans for food or beverage, etc., and the inspection of the flanks and bent areas for forming failures. While the T-bend test can be done with manually operated instruments, the deep drawing is generally done with a hydraulic machine. This ensures continuous and reproducible deformation, and also allows for measuring of the actual force that is applied.

Shock deformation is tested by impact tests with a ball-tipped weight falling from a defined height onto the test surface. The energy that can be applied before the paint delaminates is denoted (EN 13523-5; EN ISO 6272). The values are traditionally given as fall height · weight, i.e. inchpound (inlb), but, more properly, they are converted in Joules (J), using the official metric SI (Système International d´Unités) unit as required by European law [7].

7.3.2.2 *Other adhesion tests*

The cross-hatch test comprises the application of two perpendicular sets of parallel cuts on the coated surface. The cuts are spaced by 1 mm (2 mm in the case of coatings above 60 μm thickness), thereby forming a grid of small squares in the intersecting area. Adhesion is assessed by the number of squares that are removed by an adhesive tape. According to standard EN ISO 2409, the failure is classified by a code, Gt0 to Gt5.

Further adhesion tests comprise peel (e.g. ASTM D-1876-08) [8], stone chip and "coin scratch" methods, as well as tests regarding foil adhesion.

Information about the adhesion between primer and topcoat (intercoat adhesion) can also be obtained by the MEK rub test. This is carried out by rubbing the coated surface with a cloth or cotton wool ball that is soaked with methyl ethyl ketone (2-butanone, MEK). The soaked fabric is moved up and down the surface under a load, and the number of strokes counted it takes to soften and remove the paint. In general, the test shows whether the coating is properly cured. However, close inspection of the failure area can also reveal differences in the solvent stability between primer and topcoat layer. Also, it is possible to identify the removal of the topcoat from an intact primer coating, hence recognise poor intercoat adhesion. The test can be performed manually, but more preferably because of the better reproducibility, it is carried out using a machine with an oscillating probe head that is loaded with a defined mass weight, e.g. 1 kg. A number of 100 double rubs is commonly accepted as indicating sufficient curing (EN 13523-11, cf. Chapter 10.5.4.3; see also Chapter 5.5.5.2).

Table 7.3: Inspection of adhesion without deformation

- **Simulation tests without deformation**
  - **Cross-hatch:** Cutting of a square grid (1 · 1 or 2 · 2 mm²), taping and assessment of delaminated squares
  - **Other tests:** Peel tests, stone chipping, coin scratch etc.
- **Thermo- or sonographic methods**
  - Propagation of reflection patterns of heat or ultrasonic waves for the detection of failures; no numerical values of adherence

Thermographic or ultrasonic echo methods invented for thickness measurement can also be used to detect adhesion failures by analysis of changes in the propagation and reflection of heat or sound in the coatings, however do not render numerical values for adhesion strength.

7.3.3 *Laboratory methods*

7.3.3.1 *Atomic force microscopy*

Topographic and surface force data in a sub-microscopic scale can be obtained by atomic force spectroscopy, AFM. On this scale, van-der-Waals attraction and repulsion forces are measured that take effect between a surface and an approaching or retreating probe. In the apparatus, the probe is mounted on a cantilever tip that is moved closely over the studied surface, responding to the attraction or repulsion by a deflection from its original position. Monitoring of the cantilever motion renders force-distance diagrams. By scanning of a surface area with an oscillating cantilever, a true topographic chart is obtained. Due to its applicability in gaseous and liquid environments, the method can be used to study the adhesion phenomena on metallic or oxidic surfaces that are covered with an electrolyte. Several modes of measurement can be carried out, including topographic mapping, microadhesion assessment, microhardness by indentation of the cantilever tip, and the chemical interaction between the surface and the cantilever tip that, in turn, can be modified by chemical coating. AFM can also be used to measure the deflection of a stressed material. The very versatile and sensitive method even enables electric field and pH changes in the electrolyte to be measured, or adhesion investigations on single molecules [9].

Table 7.4: Principle and applications of Atomic Force Microscopy

Atomic Force Microscopy, AFM
Displacement of a cantilever tip upon approach to and retraction from a surface (van-der-Waals forces) Very high accuracy down to molecular scale
Applications: Force-distance diagrams, adhesion phenomena on electrolyte-covered metallic or oxidic surfaces, effect of chemical anchor groups, effect of chemical changes of the electrolyte, topographic mapping (with oscillating cantilever), measurement of profile (indentations), micro-roughness, and micro-hardness

7.3.3.2 Dynamic mechanical analysis

As discussed already in Chapter 5.3.2, thermosetting (duroplastic) materials are characterised by their glass transition temperature T_g that indicates the change of state between two regimes of elastic deformation with a different elasticity (storage modulus). Below T_g, the polymer is in a glassy, rigid state with a high storage modulus, and hence only little formability. Actually, most polymers used in coil coating paints are relatively brittle below this temperature. At T_g, the short-range order, caused by intermolecular forces, e.g. van-der-Waals forces, dipole interaction or hydrogen bonds, is broken up, giving way to viscoelastic behaviour characterised by an often several hundred times smaller modulus. A coated sheet therefore should be shaped at temperatures above T_g. Hence, a material's T_g is a very useful thing to know, and this information can be obtained by dynamic mechanical analysis, DMA. The method is also sensitive enough to allow detection of smaller transitions, e.g. side chain motions in a polymer molecule. Moreover, DMA gives insight into other related properties like the crosslink density or the dispersity of a polymer, the degree of copolymerisation, or the resistance of a polymer film vs. impact or scratch.

DMA experiments are carried out by exerting a harmonically oscillating stress (tensile, twisting or shearing force) on the material and monitoring the responding strain (elongation, displacement). In real polymeric systems, the strain response will lag behind the applied stress, as some of the energy is consumed and dissipated by displacement of sections of the polymer. In a diagram versus time, the oscillating strain curve will therefore be shifted by a certain phase angle from the corresponding stress curve. The phase angle is a characteristic of any polymeric material, indicating its viscoelastic behaviour. The lost energy is usually expressed through the loss modulus, while the elastic part is given by the storage modulus.

In common practice, a paint film (free film or film spread on a thin glass or carbon fibre support) is mounted between the clamps of the DMA head, and slightly elongated, to ensure that the sample is under tension throughout the experiment. A small, maximum strain (elongation) is selected (usually 0.1 %), and a sinusoidal stress then applied at a preset frequency, its amplitude being controlled versus the obtained strain, the actual motion being measured by an induction coil system. During measurement, the sample is then subjected to a steady, slow temperature increase (1 K/min).

It must be noted that both the oscillation frequency and the heating rate do influence the strain level that is measured. Also, pigments and fillers can have a reinforcing effect that shows up in an increased storage modulus, more perceptible in the rubbery regime.

The T_g can be defined in different ways, e.g. assigning it to the onset or midpoint of the transition. For documentation, it is required to state the experimental method, stress frequency and temperature ramp values, as well as the determination method of the T_g. According to standard ASTM E 1640-09, T_g is defined as the temperature at the intersection of the tangents applied, in a linearised modulus versus temperature diagram, to the storage modulus' high plateau before the transition and to the inflection point of the slope during the transition. This American standard was transferred into an international standard in 2012 (ISO 6721-11, cf. Chapter 10.5.4.4), and is likely to be adopted by CEN soon, as other parts of this standard family [10–13].

7.4 Corrosion testing

7.4.1 Outdoor exposure tests

The anti-corrosive performance of coil coatings determines the suitability of the work piece that is finally made of the coated sheet metal for the various environments it is going to meet during its service life. In order to obtain a good estimation about the durability that is to be expected, in addition to the testing of general properties and the adhesion and formability features, coated sheet specimens are subjected to a variety of different corrosion tests.

Whenever coated metal is used outdoors, as most coil coated sheet does, keeping in mind that the overwhelming part of it is used in architectural or transport (caravan, coaches, trailers, containers etc.) applications, weathering in a natural environment must be considered. Here, the metal and its paint coat meet solar irradiation, oxygen, moisture, temperature and polluting, natural and man-made, agents. The first two mentioned are ubiquitous on earth; hence they play the most important role in all degradation processes of painted goods. Pollution can be very high in places, like in industrial and urban areas, while in the countryside, it is generally low. Humidity is high at sea or in tropical regions, but low in arid landscapes. The entire set of weathering conditions is subject to diurnal and annual variations [14].

A general standard, EN ISO 12944-2, codifies climate and corrosivity conditions, to allow an overall performance classification of coatings accordingly, e.g. C2 = low; C3 = moderate; C5 = very strong (cf. Chapter 2). In principle this classification is based on weight loss rates on steel and galvanised samples. The corrosivity class of a certain site need not necessarily be the same forever, but may change from year to year; hence data logging is useful to do.

Most plastics and paints are tested in outdoor exposure, covering various climates with regard to solarisation, temperature and humidity. The test procedure is codified in the

Figure 7.3: Outdoor exposure sites; a) site of the Swedish Corrosion & Metals Research Institute, KIMAB, in Kvarnvik, Bohus Malmön, Sweden. The site is located on the west coast of the offshore island in the Skagerak, with a typical north-west European marine climate, classified C3-5 for steel, C3 for zinc (aggressive marine); b) urban test site in the centre of Prague, Czech Republic. The climate in this 1.5 million habitation area is dominated by street traffic and NO_x pollution; classification is C2 for both substrates. The facility is managed by SVÚOM, a Czech company specialising in corrosion testing

Source: Institut de la Corrosion/French Corrosion Institute

Table 7.5: Selected outdoor exposure test sites relevant for coil coating

General	
Corrosivity classification	EN ISO 12944-2
Outdoor Exposure	EN 13523-19
• **ECCA Eurodes Programme**	
- Brest (France); aggressive marine; corrosivity class C5 (steel)/C3 (zinc)	
- Geleen (Netherlands); continental/industrial; (C2/C2)	
- Hook of Holland (Netherlands); marine/industrial; (C3/C2)	
- Lisbon (Portugal); high UV exposure; (C2/C2)	
• **Battelle Institute Testing Facility**	
- Daytona Beach, Florida, USA: 29° northern latitude, high solarisation	

standard EN 13523-19. Exposure sites are chosen to cover the climate conditions that are relevant for the intended use of the coated metal. Hence, for its own corrosion test programme EURODES, the European Coil Coating Association has certified several test sites in Europe, to meet the above consideration. Locations in Brest (France), Geleen (Netherlands), Hook of Holland (Netherlands), and Lisbon (Portugal) offer the required range of conditions regarding solar irradiation, temperature, humidity and atmospheric contamination (see Table 7.5) [15]. Further regional facilities in Europe are located in e.g. Bohus Malmön (Sweden), Fusina (Italy), Hendaye (Southern France), Prague, (Czech Republic), or Sandwich (Great Britain) [16, 17]. Figure 7.3 shows the facilities in Bohus Malmön (aggressive marine) and Prague (urban).

Test panels are prepared with standardised deformations and damages (scribe, scratch), and positioned at defined declinations, to simulate the intended application and its typical load. For the European sites, these positions are 5° S (low pitched roofing), 90° N (cladding), and 45° S (solar exposure). The test duration in the European exposure sites is two to three years.

The Battelle Institute's test facility near Daytona Beach, FL, USA (29° N) is frequently used also by European paint and plastics manufacturers and users. It offers a very high solarisation and saline air, as well as subtropical humidity. Panels are usually mounted in racks at 5° and 45° S declination, and tested for up to five years [18].

Though outdoor exposure is indispensable in order to obtain a realistic picture, its major drawback obviously is the long test duration. To accelerate the corrosion progress, the test specimens may be sprayed weekly with NaCl solution. The acceleration factor is usually three to four, so that results are available within half a year. In the automotive industry, combined test programmes are common that include laboratory testing for humidity and salt spray, as well as exposure in the driving range [19].

7.4.2 Accelerated corrosion tests

7.4.2.1 Overview

To enable forecasts of the anticorrosive performance of a coating, a number of accelerated corrosion tests have been designed that simulate a long-term service life in a time lapse. Several salt spray and humidity resistance tests can be performed either at constant climate conditions, or they can be combined with repetitive cyclic variations in temperature, humidity, additional chemical or radiation load, etc. [20-23].

Typical accelerated corrosion tests include salt spray, boiling and condensation tests, either in a constant or a cyclic climate. Photochemical degradation of coatings that leads to chalking, discoloration, and gloss reduction is tested in combined humidity and irradiation tests (visible light and ultraviolet, UV). Specialist tests apply for sheet metal that is used for

Table 7.6: Corrosion tests relevant for precoated metal

Data acc. to Meuthen, Jandel[6], Kittel, Streitberger[21], Goldschmidt, Streitberger[22], Le Bozec et al.[23]

Test, Feature	Standard
Constant Climate Tests	
Salt spray – neutral	EN ISO 9227 NSS (DIN 50021), ASTM B117, EN 13523-8
Salt spray – acetic acid	EN ISO 9227 AASS
Salt spray – copper acetate	EN ISO 9227 CASS
Humidity	EN 13523-25 (DIN 50017 KK)
Condensation	EN ISO 6270-1 and 2, EN 13523-26
Filiform	EN ISO 4623-2, EN ISO 4628-10
Boiling, water soak	EN 13523-9
Cyclic Climate Tests	
Humidity/temperature cycle tests - Humidity - Tropical test - Humidity-freeze test	 EN 13523-25 (DIN 50017 KFW) analogous, ambient to 40°C analogous, - 5°C to ambient
Humidity in SO_2 atmosphere	EN 13523-13 (DIN 50018)
Cyclic salt spray/humidity tests - Automotive industry - Prohesion test	 EN ISO 11997-1 (VDA 621-415; future: SEP 1850, VDA 233-102); Renault ECC1 D172028, Volvo VCS1027.149 VW PV1210, other OEM specifications ASTM G85, A5
Cyclic Immersion test	ISO 15710, EN 3212
Weathering tests	Xenon Test, VDA 612-430 etc.

appliance manufacture. Here, the resistance to water soak, resistance to domestic cleaners, foodstuffs, dyes and colorants etc. is important to assess. Can coatings e.g. for beverage cans, lid and tabstock, and food containers, have to undergo additional tests on porosity (iron exposure value or copper spot test with acidulated copper salt solution) and sterilisability or retortability (pressurised cooker). Table 7.6 summarises important accelerated corrosion tests.

7.4.2.2 Salt spray test

General information about the overall corrosion resistance of coated metal is usually obtained in salt spray testing. For preparation, the coated specimen is damaged with a standardised scribe stylus. Pinpoint, rectangular and U-shaped cutting tips are known. As the shape of the scribe determines the area of bare metal that is exposed, and hence influences the onset and progress of corrosion events, agreement must be made as to which scribe geometry is chosen. In coil coating, the rectangular tip with a width of 1 mm (model according to Sikkens) is most commonly used. To assess possible influences on the test result that may be given by the orientation relative to the strip (rolling direction), it is usual to apply two scribes, oriented either perpendicularly to each other or in the arrangement of a St. Andrew's cross. In the case of galvanised steel, the scratch, by convention, shall not penetrate through the zinc layer, so that the basic steel substrate is left unexposed. Uncoated areas of the test panel, i.e. bare reverse faces and edges, are then covered, which is usually done with adhesive foil or tape. The specimens are then placed in a compartment where they are sprayed, at a constant temperature of usually 35 °C, with a salt solution that contains 50 g/l sodium chloride (NaCl). Steel and galvanised steel substrates are usually tested in a neutral salt fog, while for aluminium (Al) the salt solution is acidulated with acetic acid (pH ≈ 3.2, AASS Test). After a test period that is subject to agreement, the specimens are taken from the test compartment and inspected for corrosion. Paint detachment from the scribe and, when applicable, from a freshly cut edge (creepage) is measured in millimetres, and blistering on the panel area is observed and denoted.

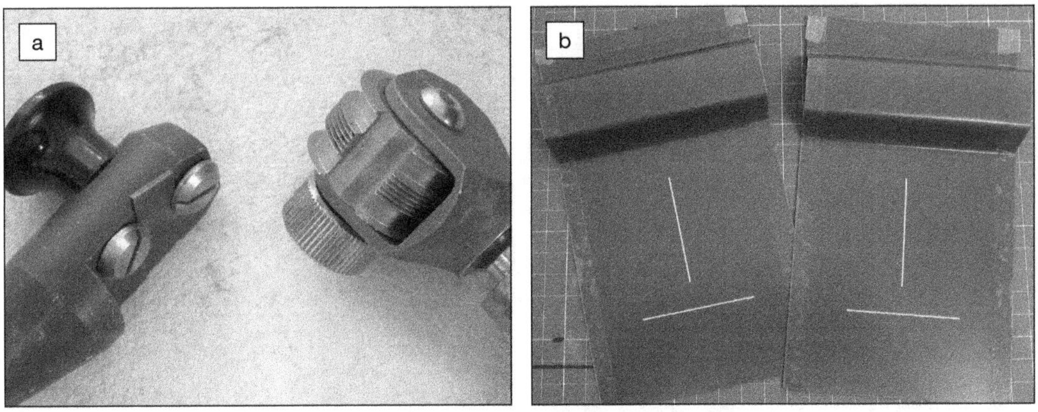

Figure 7.4: Preparation for accelerated testing; a) tools for applying test scratches; left: Scribe stylus according to Sikkens; right: Multiple cutting tool for the generation of a 1 · 1 mm² cross-hatch; there are also tools with 11 parallel edges available; b) panels prepared with two perpendicular scribes of 1 mm width, and a standardised trapezoidal profile deformation; the reverse side and three edges are covered with adhesive foil

The panels may also have a deformation that again is subject to the agreed specification. The deformation can be applied along or across the specimen. Triangular, rectangular, or trapezoidal profile deformations, in some cases with varying bending radius, are common.

For steel substrates, typically 336 to 1,008 h (2 to 6 weeks) test time is required for steel and galvanised steel, after which period only small defects are accepted. Aluminium substrates are usually tested for extended duration, i.e. 1,512 to 2,016 h, due to the intrinsically higher corrosion resistance of this metal. Al sheet for aircraft meets the harshest requirements, in that, after 3,024 h, a creepage of merely 1.25 mm from the scribe is accepted.

7.4.2.3 Constant climate humidity

The uptake of water or moisture generally leads to swelling and plastification of coatings. Accumulation of water may weaken bonds at the interfaces in a coating, causing hydrolytic cleavages. Finally, also chemical degradation of the polymer film and corrosion of the metal substrate can be promoted [24]. It is therefore important to assess the resistance of coatings against humid conditions. A number of tests are related to this task, each focusing a particular humidity environment. In the constant climate humidity test, panels that are prepared in a similar way as for salt spray are stored in a compartment under constant humidity and temperature. Typical conditions are 40 °C and > 85 % relative humidity (rH). Inspection after an agreed test time is done regarding blistering and other aspect changes, like corrosion

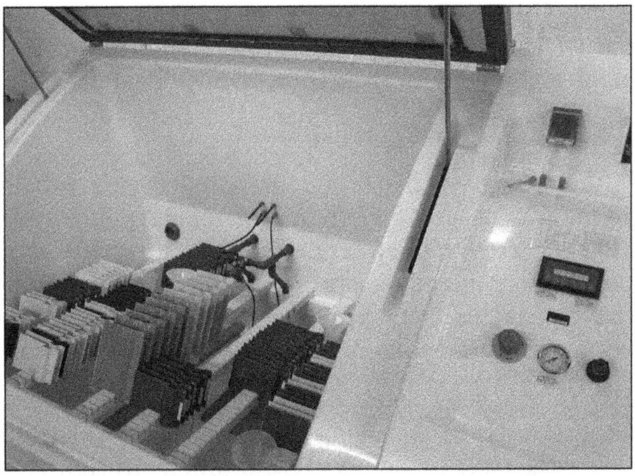

Figure 7.5: Opened corrosion test chamber; the compartment can be used for humidity (condensation) and salt spray tests, with adjusted electrolytes, atmospheres, and programmable temperature and relative humidity; similar test compartments are available for time programmed cyclic testing

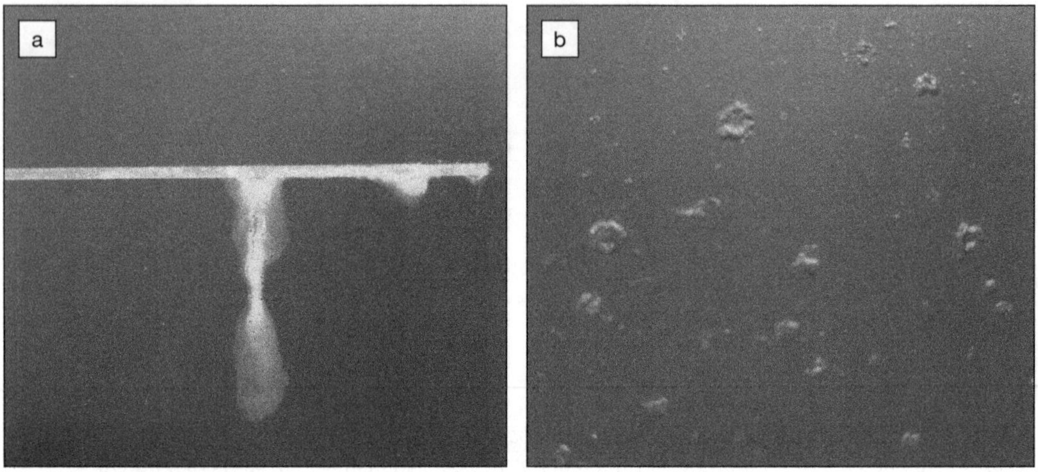

Figure 7.6: Degradation of a developmental polyester coating on galvanised steel by accelerated testing; a) creepage and white rust at the scribe after neutral salt spray (1,008 h); b) blisters after humidity test (1,008 h)

products, colour or gloss changes, etc. The duration of the tests is usually chosen as for the salt spray, e.g. 2 to 6 weeks for steel substrates incl. galvanised steel.

7.4.2.4 Condensation

The resistance of coatings against increased temperature and permanent humidity is assessed in condensation tests (QCT Test, EN 13523-26). The test panels are stored at condensation conditions (i.e. rH close to 100 %) and fixed temperature, e.g. 40 or 60 °C. After the test (e.g. 1,000 or 1,500 h), they are checked for blisters and other visible defects [25].

7.4.2.5 Water soak and boiling tests

Water soak tests are usually carried out by storage of the test specimens in deionised water at 40 °C. However, for coil coated metal, the temperature is commonly increased to boiling, and the test is aggravated by applying a deformation by bending or cupping, the latter optionally even combined with a crosshairs or cross cut. Adhesion losses, if applicable, in the deformed area, are assessed by taping, and other visible changes denoted. The duration of the test is subject to agreement. Specifications may range from 15 min to several hours.

7.4.2.6 Filiform corrosion

Filiform corrosion displays a threadlike corrosion pattern that is seen, especially on coated aluminium, in chloride containing, reduced-humidity (approx. 80 % rH) environments. This kind of corrosion often occurs underneath coatings with a high barrier effect vs. penetration by oxygen and moisture. It is characterised by active anodic dissolution of the metal that propagates, along grain boundaries of the metal, at the head of the filament, while the oxygen reduction takes place in the tail section [26, 27].

The filiform corrosion test was first invented for the aerospace sector on aluminium substrate (Lockheed Test), but has generally been adopted. It involves an initial 1-hour incubation of the test specimens over fuming hydrochloric acid (HCl 36 %). Subsequently, panels are held under constant atmosphere with controlled humidity and temperature (82 % rH, 40 °C). After the agreed test period, e.g. 1,008 h, the specimens are inspected for the number and length of threads that have developed. Steel substrates are incubated with NaCl solution rather than HCl vapour (EN ISO 4623-1 and -2; EN ISO 4628-10).

7.4.2.7 Cyclic humidity

Cyclic tests have been invented in order to obtain a more realistic response than in a constant climate. Natural exposure goes along with daily and seasonal changes which has an effect on the progress of corrosion and degradation. The changes may cause an aggravation, like temperature extremes, but they may also lead to relaxation, e.g. when the test specimen is allowed to dry which usually slows down degradation reactions. Repetitive cyclic variations are programmed for temperature and humidity. Often, periods with chemical or radiation stresses are integrated in the test cycle.

Cyclic humidity and condensation tests are usually carried out at the appropriate relative humidity, changing the temperature between a high and a low value. Tropical tests with temperature cycles between ambient and 40 °C, or freezing and thawing tests are common. The cyclic humidity test is also carried out in a sulphur dioxide containing atmosphere (SO_2, EN 13523-13), to simulate “acid rain”. The required duration and the number of cycles undergone during a test are subject to agreement.

7.4.2.8 Prohesion

Though it continues to be used as the most common standard test, the neutral salt spray has proven to be too harsh on galvanised steel. Real corrosion events in a natural environment generally do not result in the amount of deterioration of the galvanised surfaces as does the laboratory test. To obtain a more realistic picture, the prohesion test was invented. It comprises cycles of a slightly acidulated salt spray with a diluted NaCl solution, and dry intervals. The typical salt solution contains 3.5 g/l ammonium sulphate, $(NH_4)_2SO_4$, and 0.5 g/l NaCl in deionised water. First designed for steel substrates, the test has become common for aluminium, as well. Again, Al sheet for the aerospace industry faces the highest requirements, i.e. up to 2,016 h (12 weeks) [28].

Another similar test procedure is discussed particularly for aluminium substrates, in order to improve the comparability of results between the acetic acid salt spray (cf. Chapter 7.4.2.2) and outdoor exposure. The artificial test renders blistering, while the natural weathering in a marine environment leads to filiform corrosion. The proposed test involves a repetitive 8-hour cycle of acetic acid salt spray, drying at 50 °C and extended storage at 40 °C and 80 % rH. The corrosion pattern, when compared with outdoor exposure results (Bohus Malmön), has been found to correlate well [29].

7.4.2.9 VDA test

Standardised cyclic tests are particularly common in the automotive industry, in order to apply conditions that are in better accordance with the reality of a car’s lifetime. The relevant standard 621-415 of the German Automobile Association (VDA) is generally accepted as a guideline. EN ISO 11997-1, Annex D, embraces the conditions of the VDA test cycle [30] that involves a succession of salt spray, humidity and open-air storage periods within a week. Nevertheless, individual OEMs like Daimler, Ford, General Motors, Renault, Volkswagen, or Volvo have issued internal test specifications that deviate from the VDA cycle regarding the order and duration of load and relief periods [31].

Usually, a minimum of 10 cycles (10 weeks) is specified as test duration for cathodic electro-dip coats, the general basic coating on a car body. Testing of coil-applied corrosion protection primers hence is normally done with an electro-coat applied over the primer. The entire paint finish, usually composed of a filler, a (coloured) basecoat and a clearcoat layer, must endure longer terms.

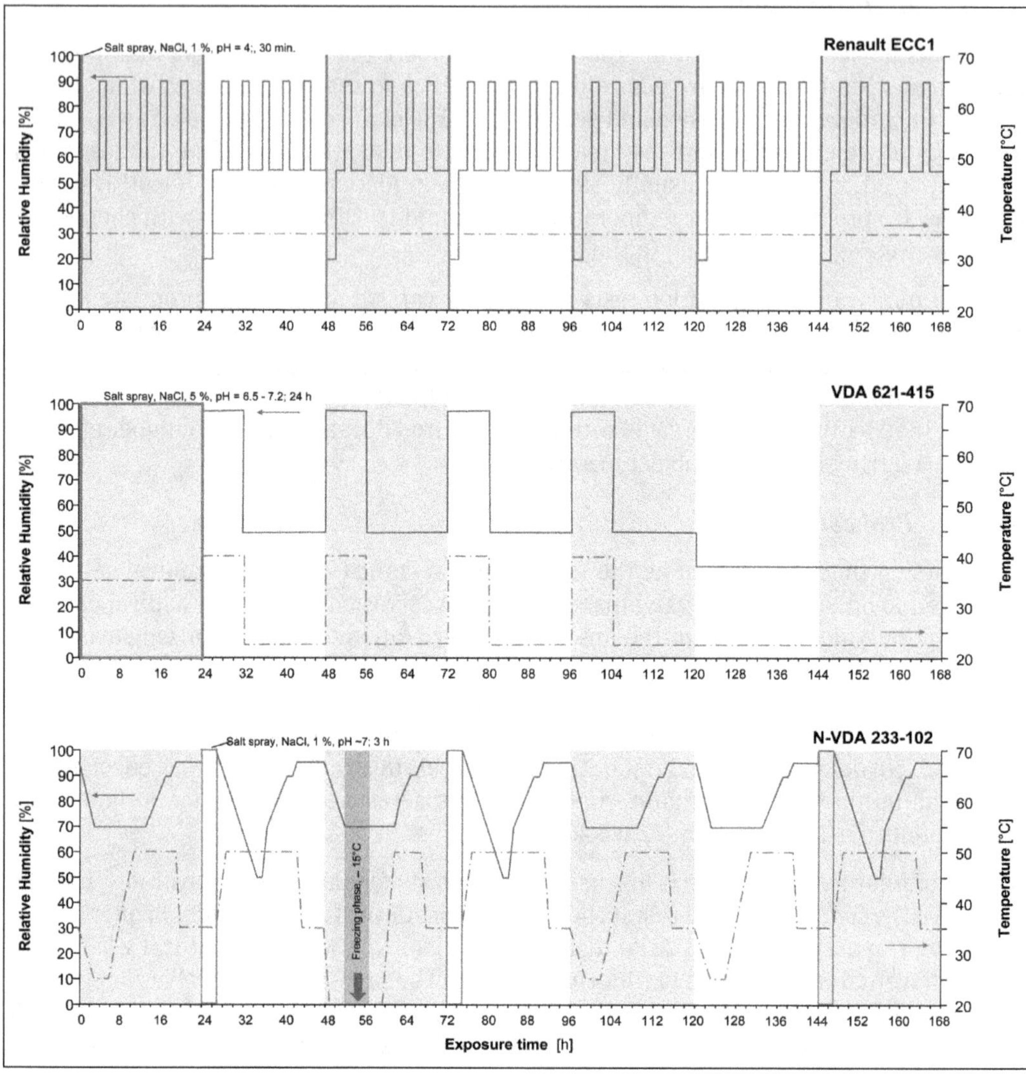

Figure 7.7: Comparison of selected cyclic test schedules in the automotive industry. The new, complex test cycle VDA 233-102 (bottom) is composed of three individual 24-h sub-cycles that can be arranged in different sequences. One sub-cycle includes a 5-h freezing phase Source: Thierry[31] and Prosek et al.[32]

To further optimise the comparability between accelerated tests and the reality, there is an ongoing discussion and experimental work about how a laboratory test cycle must be adapted to fulfil this purpose. Hence, a new test procedure (N-VDA test, VDA 233-102) has been developed and put in a standard [32, 33]. It will supersede the old standard, however has not officially become effective yet. Figure 7.7 displays the schedules of the Renault ECC1, and both old and new VDA tests.

7.4.2.10 UV test and weathering

The resistance of the binder system against photochemical degradation is assessed by exposing the coated specimens to ultraviolet irradiation (UV). As temperature and humidity play an important role in photooxidation and light-induced hydrolysis processes, as occur during natural weathering, the UV exposure is combined with humidity (QUV Test, EN ISO 11997-2, EN 13523-10, VDA 621-430). The radiation can be chosen to simulate solarisation with the entire UV/Vis spectrum (Xenon lamps), or special UV lamps are used that emit a smaller spectral segment. Commonly, UV-A (maximum at 340 nm) or UV-B lamps (313 nm) are used. The latter radiation represents a climate with extreme solarisation. The energy at this wavelength is sufficient to initiate photooxidation by cleavage of any chemical bond in polymers [34]. Up to 2,000 h test time or more are usually required, after which period the specimens are inspected for discoloration (chalking), adhesion losses or traces of corrosion products. Though primarily addressing the performance of the binder resin with this test, usually a representative weatherability picture of the entire coating system is obtained. Test equipment is pictured in Figure 7.8.

Figure 7.8: Two generations of QUV testing compartments. The machines are programmed for alternating UV irradiation and condensation, e.g. each period lasting 4 h at a preset temperature, usually between 30 and 60 °C

7.5 Electrochemical testing

7.5.1 General remarks

While outdoor weathering and accelerated testing provide a good account on the aspect of corrosive degradation, corrosion science employs a wide variety of electrochemical methods in order to gain a better understanding of corrosion events and the associated delamination of paints [35, 36]. Experimental designs are directed towards monitoring and prediction of corrosion, as well as to tailoring anticorrosive strategies [37]. Electrochemical testing is being combined with complementary analytical methods, e.g. optical (surface-enhanced and total-reflection infrared and Raman methods), electron-optical and atomic force and synchrotron

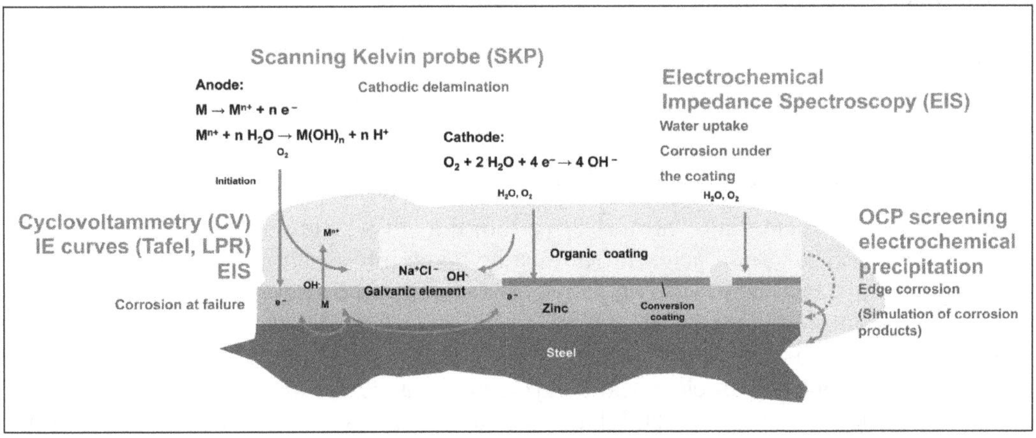

Figure 7.9: Electrochemical events at a corroding metal/coating interface and related investigation methods

Illustration acc. to Keil[43]

Table 7.7: Standard electrochemical potentials of selected elements and compounds oxidation occurs more readily with lower potential

Description	**Electrode reaction** in aqueous solution	**Standard potential** E_0 at 25 °C (V)
Pt | F_2 | F–	F_2 (g) + 2 e^- → 2 F^-	2.87
Au | $AuCl_4$ $^-$, Cl^-	$AuCl_4^-$ + 3 e^- → Au + 4 Cl^-	1.00
Ag | Ag^+	Ag^+ + e^- → Ag	0.7991
Pt | Fe^{2+}, Fe^{3+}	Fe^{3+} + e^- → Fe^{2+}	0.771
Pt | H_2 | H^+	2 H^+ + 2 e^- → H_2	0.0000
Pb | Pb^{2+}	Pb^{2+} + 2 e^- → Pb	-0.126
Fe | Fe^{2+}	Fe^{2+} + 2 e^- → Fe	-0.4402
Zn | Zn^{2+}	Zn^{2+} + 2 e^- → Zn	-0.7628
Al | Al^{3+}	Al^{3+} + 3 e^- → Al	-1.662
Mg | Mg^{2+}	Mg^{2+} + 2 e^- → Mg	-2.363
Pt | MnO_2 | MnO_4^-	MnO_4^- + 2 H_2O + 3 e^- → MnO_2 + 4 OH^-	0.588
Pt | O_2 | OH^-	O_2 + 2 H_2O + 4 e^- → 4 OH^-	0.401
Pt | H_2 | OH^-	2 H_2O + 2 e^- → H_2 + 2 OH^-	-0.8280

According to Barrow [8]

radiation methods [38-41]. When suit- ably designed, experiments even allow for the real-time investigation of corrosion and delamination events [42]. Figure 7.9 depicts various electrochemical methods and the degradation events they target [43].

7.5.2 Electrochemical potential

7.5.2.1 Standard potential

Electrochemical reactions occur via the exchange of electrons between chemical species that thereby change between reduced and oxidised states. This kind of reactions, hence called red-ox reactions, is associated with a characteristic voltage depending on the chemicals involved. When measured by a galvanostatic method versus a hydrogen electrode at standard concentration and temperature (i.e. its potential defined as 0 V), the potential of a red-ox system is called a standard electrochemical potential. In particular, it describes how easily metals are oxidised. Negative standard potentials indicate a metal that is prone to oxidation (hence corrodes fast), while a positive characterises a metal that is resistant. At the upper end of a standard potential chart (electrochemical series), we find the so-called noble metals. Table 7.7 shows selected standard potentials [44].

7.5.2.2 Corrosion potential/current monitoring

While the standard potential gives a relative ranking of the oxidisability of metals, it does not describe a real corrosion potential. It must be taken into account that the actual potentials and currents associated with an active corrosion event in fact change while the reaction proceeds. Hence it is not possible to measure either factor directly and independently. Suitable experimental setups are required that allow interpolation of the limit values for both factors. By applying an electrical polarisation by a step or sweep voltage increase in the proximity of its corrosion potential, a corrosion system is moved out of its equilibrium. The oxidative and the reductive branch can be assessed separately, measuring the finite response current. Oxidation takes place at positive, reduction at negative currents. Suitable

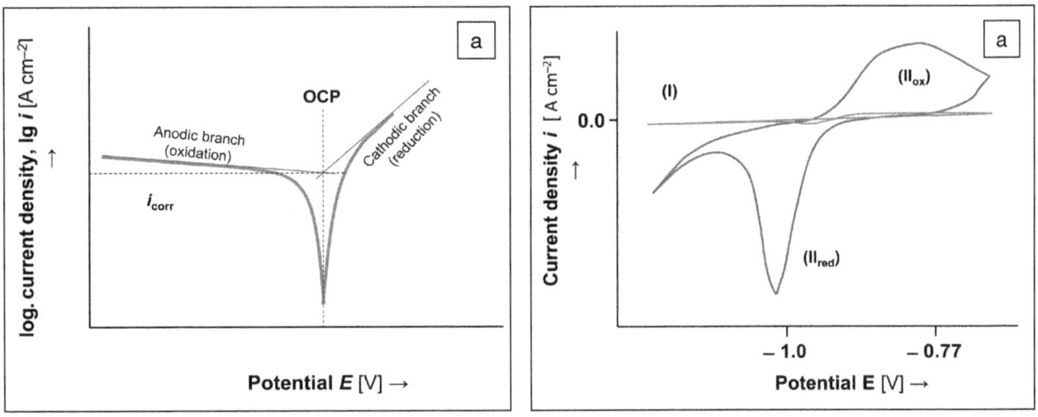

Figure 7.10: Potential/current measurements to investigate corrosion; a) schematic representation of a Tafel plot in an lg i-E diagram; b) cyclovoltammograms on galvanised surfaces; (I) native oxide on the Zn surface inhibits red-ox reactions resulting in near-zero current; (II) metallic Zn, freshly etched by treatment with an alkaline solution, is reactive enough to undergo oxidation II_{ox}, while its dissolved ions become available for the inverse reductive reaction II_{red} *Illustration a) acc. to Mansfeld [45]; b) acc. to Fink et al. [47]*

plotting of the potential vs. the current allows assessing the corrosion or **open circuit potential**, OCP, versus the zero or minimum current condition. Often, a plot of the potential E vs. the logarithm of the responding current density, lg $\boldsymbol{i}$, is chosen. Figure 7.10a shows such a plot (**Tafel plot**).

Furthermore, a normalised plot of $\boldsymbol{i}/\boldsymbol{i}_{corr}$ vs. (E - OCP) in the vicinity of the OCP (approx. ± 15 mV) gives information about the **linear polarisation resistance**, LPR. This feature can be interpreted as the kinetic hindrance that is faced by ions being dissolved from the metal, when passing through the double layer and migrating away from the surface. The plot renders a straight line, its slope representing the associated resistance [45].

A dynamic method to investigate red-ox systems, the so-called **cyclovoltammetry**, CV, is done by measuring the potential around the OCP and the resulting current density by a cyclic sweep of the voltage. Again, both red-ox events can be monitored, when the voltage is only changed for small values. In an E-$\boldsymbol{i}$ plot, a closed-loop curve is obtained. The area enveloped by the curves, above or under the zero-current line, is a measure of the reaction turnover. When the reaction is entirely reversible, the closed loop will be passed through over again as often as the voltage sweep is repeated. In case of irreversible systems, e.g. when onward reactions remove components from the equilibrium, transitory curves are obtained, and the equilibrium is shifted. An exemplary voltammogram on a technical zinc (Zn) surface is depicted in Figure 7.10b [46, 47].

7.5.3 Electrochemical impedance spectroscopy

Corrosion and degradation reactions in the metal-to-paint interface as well as in the bulk paint bring about an electrochemical activity that can be measured by means of electrochemical impedance spectroscopy, EIS [48]. A coating forms a barrier to migrating ions and flowing electrons alike, and shows both resistance and capacitance properties when subjected to a sinusoidal alternating voltage. The behaviour of the coating can be simulated by an equivalent circuit. An intact coating is represented by a simple circuit containing a series composed of a resistor and a loop with a second resistor and a capacitor in parallel [49]. The first resistor simulates the ion transport through the electrolyte which is slower than free flowing electrons, while the latter is identified with ion transport through pores and

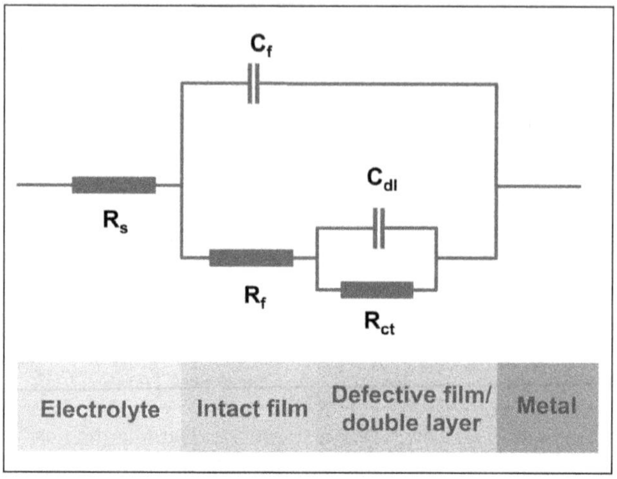

Figure 7.11: Schematic: EIS equivalent circuit; R_s = resistance of solution, R_f = film (pore) resistance, R_{ct} = charge-transfer resistance (electron transport through interface); C_f = film capacitance, C_{dl} = double-layer capacitance

Illustration acc. to Lewis [55]

interstices in the coating (pore resistance). The capacitance resembles the dielectric features of the coating, as it separates the metal from the electrolyte. With the electrochemical activity associated with corrosive events at the metal end of the pores, more layers emerge that act like additional components in the equivalent circuit, and can be assigned to double layer capacitance, and the polarisation resistance due to charge transfer [50]. Figure 7.11 gives a schematic illustration.

An alternating voltage, oscillating around the open circuit potential (OCP) with a small amplitude, e.g. ± 10 to 20 mV, applied to a coated test specimen, will result in a response current that alternates at a certain phase angle depending on the resistor and capacitor values the coating takes on. Performing a series of measurements at different frequencies renders a set of data which can be used for calculation of the total impedance (**Z**, measured in Ωcm^{-2}) and the phase angle (Θ). Separating **Z** into real ($\mathbf{Z}_{re}$) and imaginary components ($\mathbf{Z}_{Im}$) allows plotting these against each other in a diagram (Nyquist plot). Another representation of the results is the double logarithmic plot of the absolute value (or modulus) of the impedance, log $|\mathbf{Z}|$, or the phase angle Θ vs. the frequency (Bode plots). Shifts and deformations in the plots of repetitive EIS cycles indicate corrosive events. At low frequencies, i.e. $\leq$ 0.1 Hz, the impedance resembles the resistance of the coating to direct current.

An ideal (insulator) barrier has no pores, hence an infinite pore resistance. It would therefore be represented by a serial array of the solution resistance $\mathbf{R}_s$ and the film capacitance $\mathbf{C}_f$ only (see Figure 7.11). In the Nyquist diagram, such a circuit displays only the real component as a single peak at value $\mathbf{R}_s$. The associated Bode plots show a base line at this value for high frequencies, whereas towards low frequencies the behaviour is ruled by the capacitance, demanding a linear increase of the total impedance with a slope +1. The phase angle is 90° at the low, and 0° at the high frequency end.

A real, intact coating will show a semi-circle in the Nyquist plot, the length of the base line stretching between $\mathbf{R}_s$ and the film resistance, $\mathbf{R}_f$. In this system, the Bode plot will show a constant ($\mathbf{R}_s + \mathbf{R}_f$) for low frequency that then descends to the lower level of $\mathbf{R}_s$ alone. Start and end point of this sigmoidal descent as well as its slope (at the inflection point) are determined by the film capacitance $\mathbf{C}_f$. The phase angle will rise from zero at low frequencies through a maximum at the frequency that corresponds to the inflection point mentioned above, then drop back again. Corrosive degradation renders two or more semi-circles in the Nyquist plot, depending on the number of layers that are formed, each resembling an additional resistor/capacitance circuit with its corresponding time constant. Often, instead of individual semi-circles, the enveloping curves are seen.

Repetitive frequency sweeps over time will reveal changes in the coating and hence allow the monitoring of water uptake and of the progress of corrosion [51, 52]. Usually by good

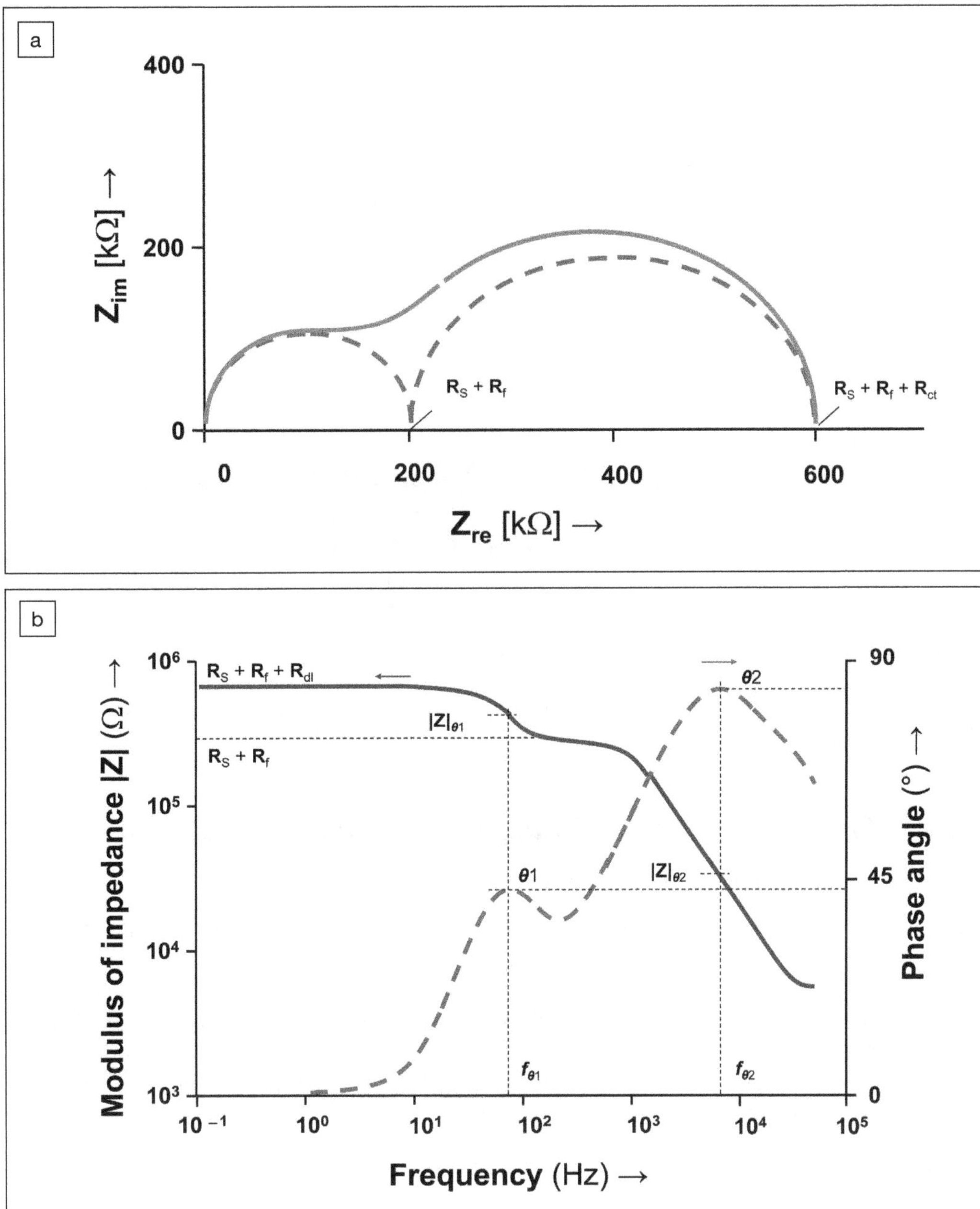

Figure 7.12: Impedance at a painted metal/solution interface; a) Nyquist plot with individual semi-circles and enveloping curve; separated semi-circles are obtained when the frequencies associated with the corrosion at the metal/coating interface and the electrolyte penetration of the paint film, and the individual pore and charge-transfer resistances, R_f and R_{ct}, are sufficiently different. Curves get even more complex when diffusion controlled processes apply for the transport of species through the pores, which has the effect of an extra serial impedance in addition to the R_{ct} value; b) Bode plot: Schematic representation (for the assignment of symbols cf. Figure 7.11) Source: Walter[53]

approximation, a value of lg |**Z**| = 8 can be considered good protection by the coating. With corrosion progress, the impedance will drop, and more complex spectra emerge. A value of lg |**Z**| ≤ 6 indicates a poor performing, degraded system [53-57].

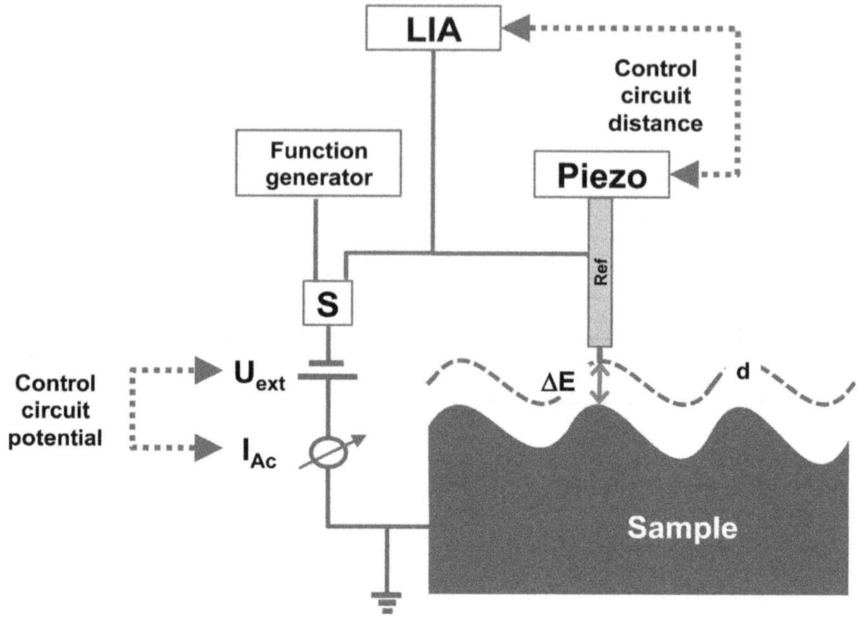

Figure 7.13: Schematic view of the height-regulated Scanning Kelvin Probe (HR-SKP)
LIA = Lock-in Amplifier; Ref = Reference Electrode with capillary tip; d = tip distance over sample
ΔE = potential difference
Source: Grundmeier, G. et al.[46]

7.5.4 Electrochemical techniques with high spatial resolution

7.5.4.1 Scanning vibrating electrode

A high-resolution chart of corrosion events at a metal/electrolyte interface can be obtained by the scanning vibrating electrode technique, SVET[58]. This method is based on the measurement of currents induced by ionic species that migrate from and to anodic and cathodic loci during corrosion on a bare metal surface that displays local inhomogeneities. With a lateral resolution in the order of some 10 μm, and a sensitivity to current densities below 5 μA/cm^2, screening of the corrosion performance of passivation treatments, cut-edge corrosion or alloy composition effects can be studied[59, 60].

7.5.4.2 Height-regulated scanning Kelvin probe

In-situ measurements of the distribution of electrical potentials of a metal surface underneath an organic coating are rendered possible by the height-regulated scanning Kelvin probe, HR-SKP technique. The method involves a reference electrode being moved closely above the surface of the test specimen to scan a small area underneath a paint film. When the coating is damaged, e.g. by a scribe, and subjected to a corrosive electrolyte to seep in, the paint film is detached due to the propagating corrosion. The SKP measures the progress of the potential, while simultaneously, a microscopic inspection of the paint film from the top is possible. Repetitive measurements reveal the propagation of the film delamination with time. By comparison of the potential and the microscopic topographic development, a distinction between corrosive (red-ox) and hydrolytic failure modes (wet delamination) can be made. Figure 7.13 shows a schematic of the experimental array of the HR-SKP. An example for a mechanistic study is described and illustrated in Chapter 4.4.4.4[61-64].

7.6 Literature

[1] Lomax, P., Trends in coating and thickness measurement, Quality digest, QCI International 2006, http://www.qualitydigest.com/aug04/articles03_article.shtml (02-08-13; 00.33 h)

[2] Bucher, U. W., Dynamic on-line wet and dry coating thickness measurement and control system, ECCA Autumn Congress, Proc., European Coil Coating Association, Brussels 2010

[3] Louis, A. K., Dörr, P., Gruss, C., Petry, H., Measurement of paint layer thickness with photothermal infrared radiometry, in: Jäger, W., Krebs, H.-J. (eds.), Mathematics - Key Technology for the Future, Springer, Berlin / Heidelberg 2003, pp. 460 ff

[4] Justice, M., Beta Backscatter for Coating Thickness Measurement: Still Viable After All These Years, Metal finishing online, Elsevier 2008; http://www.metalfinishing.com/view/4269/beta-backscatter-for-coating-thickness-measurement-still-viable-after-all-these-years/ (01-08-13; 23:55h)

[5] Souzy, S., EPSILON 5000 - On line dry or/and wet film thickness measuring system, ECCA Autumn Congress, Proc., European Coil Coating Association, Brussels 2012

[6] Meuthen, B., Jandel, A.-S., Coil Coating, 2nd ed., Vieweg, Wiesbaden 2008, pp. 189 ff

[7] anon., Council Directive 80/181/EEC of 20 December 1979 on the approximation of the laws of the Member States relating to units of measurement and on the repeal of Directive 71/354/EEC, Official Journal of the European Union, 1980, L 39, pp. 40 ff; eur-lex.europa.eu/...

[8] anon., ASTM D1876 - 08 Standard Test Method for Peel Resistance of Adhesives (T-Peel Test), Am. Soc. f. Testing and Materials (ASTM) Intern., W. Conshohocken PA, USA, 2010; www.astm.org/Standards/D1876.htm (01-08-13; 21:40 h)

[9] Grundmeier, G., Stratmann, M., Adhesion and de-adhesion mechanisms at polymer/metal interfaces: Mechanistic understanding based on in situ studies of buried interfaces, Annu. Rev. Mat. Res. 35, 2005, pp. 584 ff

[10] Lowe, C., Characterisation and measurement of polymeric materials, Vol. 1, Analytical Methods for Surface Coatings, Sita Technology, Edinburgh 2002, pp. 25 ff

[11] Lin, L., Dynamic Mechanical Analysis (DMA) - Basics and Beyond, Perkin Elmer (eds.), 2000; http://depts.washington.edu/mseuser/Equipment/RefNotes/LinLiDma-SF.pdf; (23-07-13; 13:12 h)

[12] anon., ASTM Standards for coil coating paints, National Coil Coating Association (NCCA), Chicago 2000

[13] anon., Standard test method for assignation of the glass transition temperature by dynamic mechanical analysis, ASTM, W. Conshohocken, PA, USA, 2009; www.astm.org/Standards/E1640.htm (27-07-13; 11:25 h)

[14] Schulz, U., Kurzzeitbewitterung (Accelerated weathering), F&L Edition, Vincentz Network, Hannover 2007, pp. 12 ff

[15] anon., Eurodes Programme, Eurodes: Outdoor Exposure, European Coil Coating Association, Brussels 2009; www.prepaintedmetal.eu/prg/selfware.pl?id_sitemap=190&language=EN (01-08-13; 21:41 h)

[16] Prosek, T., Field testing sites, Member Research Consortium T2 (Corrosion properties of coil-coated products), internal document, Institut de la Corrosion / French Corrosion Institute, Brest, France 2011

[17] Pietschmann, J., Gardein, R., Filiform corrosion and results of 10 years natural weathering, Galvanotechnik 106, 2008, pp. 1764 ff

[18] anon., Florida Materials Research Facility, www.bfmrf.org/ (03-08-13; 11:24 h)

[19] Goldschmidt, A., Streitberger, H.-J., Basics of Coating Technology, 2nd ed., BASF Coatings AG, Münster 2007, pp. 445 f

[20] Meuthen, B., Jandel, A.-S., Coil Coating, 2nd ed., Vieweg, Wiesbaden 2008, p. 193

[21] Kittel, H., Streitberger, H.-J., Lehrbuch der Lacke und Beschichtungen (Coursebook of Paints and Coatings), Vol. 6, 2nd ed., Hirzel, Stuttgart 2008, pp. 321 ff

[22] Goldschmidt, A., Streitberger, H.-J. [18], pp. 429 ff

[23] LeBozec, N., Blandin, N., Thierry, D., Accelerated corrosion tests in the automotive industry: A comparison of the performance towards cosmetic corrosion, Materials and Corrosion 59, 2008, pp. 889 ff

[24] Goldschmidt, A., Streitberger, H.-J. [18], p. 447

[25] Prosek, T., Nazarov, A., Stoulil, J., Thierry, D., Evaluation of the tendency to blistering: Field exposure, accelerated tests and electrochemical measurements, Symposium, Coil coated steel: Durability, functionality, and new materials, Paris, Proc., French Corrosion Institute, Brest 2008

[26] Grundmeier, G., Stratmann, M., [8], pp. 604 ff

[27] Leth-Olsen, H., Nisancioglu, K., Filiform Corrosion Morphologies on Painted Aluminum Alloy 3105 Coil Material. Corrosion 53, 1997, pp. 705 ff

[28] Yasuda, H. K., Reddy, C. M., Yu, Q. S., Deffeyes, J. E., Bierwagen, G. P., He, L., Effect of Scribing Modes on Corrosion Test Results, Corrosion 57, 2001, p. 30

[29] Bjoergum, A., Lein, J. E., Lunder, O., An alternative test method to acidified salt spray, ECCA Autumn Congress, Proc., European Coil Coating Association, Brussels 2010

[30] anon., DIN EN ISO 11997-1:2006-04, Paints and varnishes - Determination of resistance to cyclic corrosion conditions - Part 1: Wet (salt fog)/dry/humidity (ISO 11997-1:2005), German version: Beuth, Düsseldorf 2006, www.beuth.de/de/norm/din-en-iso-11997-1/87717570 (27-06-13; 21:08 h)

[31] Thierry, D., Cyclic accelerated corrosion testing - State of the art and future needs, 2nd Int'l. seminar on automotive corrosion, Swerea-KIMAB, Stockholm 2003

[32] Prosek, T., Le Bozec, N., Thierry, D., Application of automated corrosion sensors for monitoring the rate of corrosion during accelerated corrosion tests, Mat. Corros. online, DOI: 10.1002/maco.201206655, 2012; onlinelibrary.wiley.com/doi/10.1002/maco.201206655/abstract (01-08-13; 21:35 h)

[33] anon., Cyclic corrosion testing of materials and components in automotive construction, Stahl-Eisen Prüfblatt (SEP) 1850, 1st ed., (equivalent to VDA 233-102), Stahleisen eds., Düsseldorf 2012

[34] Schulz, U. [14], pp. 15; 25

[35] Prosek, T. et al., [55] pp. 92 ff

[36] Grundmeier, G., Schmidt, W., Stratmann, M., Corrosion protection by organic coatings: Electrochemical mechanism and novel methods of investigation, Electrochim. Acta 45, 2000, pp. 2515 ff

[37] Stratmann, M., Corrosion stability of polymer-coated metals - New concepts based on fundamental understanding, Corrosion 61, 2005, pp. 1115 ff

[38] Ogle, K., Tomandl, A., Meddahi, N., Wolpers, M., The alkaline stability of phosphate coatings I: ICP atomic emission spectroelectrochemistry, Corrosion Science 46, 2004, pp. 979 ff

[39] Rohwerder, M., Hornung, E., Stratmann, M., Microscopic aspects of electrochemical delamination: An SKPFM study, Electrochim. Acta, 48, 2003, pp. 1235 ff

[40] Ohman, M. , Persson, D., An integrated in situ ATR-FTIR and EIS set-up to study buried metal-polymer interfaces exposed to an electrolyte solution, Electrochim. Acta 52, 2007, pp. 5159 ff

[41] Renner, F. U., Stierle, A., Dosch, H., Kolb, D. M., Lee, T.-L., Zegenhagen, J., Initial corrosion observed on the atomic scale, Nature 439, 2006, pp. 707 ff

[42] Grundmeier, G., Stratmann, M. [9]

[43] Keil, P., private communication, BASF Coatings, Münster 2013

[44] Barrow, G. M., Physikalische Chemie (Physical Chemistry), Vol. 3, Bohmann-Vieweg, Vienna 1977, pp. 212 ff

[45] Mansfeld, F., Fundamental aspects of the polarization resistance technique - the early days, J. Solid State Electrochem. 13, 2009, pp. 515 ff

[46] Klimow, G., Fink, N., Grundmeier, G., Electrochemical Studies of the Inhibition of Cathodic Delamination of Organically Coated Galvanised Steel by Thin Conversion Films, Electrochim. Acta 53, 2007, pp. 1290 ff

[47] Fink, N., Wilson, B., Grundmeier, G., Formation of Ultra-Thin Amorphous Conversion Films on Zinc Alloy Coatings, Part 1: Composition and Reactivity of Native Oxides on ZnAl (0.05%)-Coatings, Electrochim. Acta 51, 2006, pp. 2956 ff

[48] Kendig, M. W., Jeanjaquet, S., Lumsden, J., Electrochemical Impedance of Coated Metal Undergoing Loss of Adhesion, in: Scully, J. R., Silverman, D. C., Kendig, M. W. (eds.), Electrochemical Impedance: Analysis and Interpretation, ASTM STP, publ. code: 04-011880-27, ASTM International, W. Conshohocken, PA, USA 1993, pp. 407 ff

[49] Kendig et al. [48], p. 413

[50] Murray, J. N., Electrochemical test methods for evaluating organic coatings on metals: An update. Part III: Multiple test parameter measurements, Progress in organic Coatings 31, 1997, pp. 375 ff

[51] Bonora, P. L., Deflorian, F., Fedrizzi, L., Electrochemical Impedance Spectroscopy as a tool for investigating underpaint corrosion, Electrochim. Acta 41, 1996, pp. 1073 ff

[52] van Westing, E. P. M., Ferrari, G. M., de Wit, J. H. W., The determination of coating performance with impedance measurements. 2. Water-uptake of coatings, Corrosion Science 36, 1994, pp. 957 ff

[53] Walter, G. W., A review of impedance plot methods used for corrosion performance analysis of painted metals, Corr. Science 26, 1986, pp. 681 ff

[54] Walter, G. W., A comparison of single frequency and wide frequency range impedance tests for painted metals, Corr. Science 30, 1990, pp. 617 ff

[55] Lewis, O. D., A Study of the Influence of Nanofiller Additives on the Performance of Waterborne Primer Coatings, PhD Thesis, Loughborough Univ., Loughborough 2008, p. 81

[56] Ngo, S., Tuning into corrosion, ECCA Autumn Congress, Proc., European Coil Coating Association, Brussels 2012

[57] Bierwagen, G., Tallman, D., Li, J. P., He, L., Jeffcoate, C., EIS studies of coated metals in accelerated exposure, Progress in Organic Coatings 46, 2003, pp. 149 ff.

[58] Simoes, A. M., Bastos, A. C., Ferreira, M. G., Gonzalez-Garcia, Y., Gonzalez, S., Souto, R. M., Use of SVET and SECM to study the galvanic corrosion of an iron-zinc cell, Corrosion Science 49, 2007, pp. 726 ff

[59] Thébault, F., Vuillemin, B., Oltra, R., Ogle, K., Allely, C., Investigation of self-healing mechanism on galvanized steels cut edges by coupling SVET and numerical modeling, Electrochim. Acta 53, 2008, pp. 5226 ff

[60] Taylor, C. J., Elvins, J., Sullivan, J. H., Worsley, D. A., Corrosion Performance Evaluation of Zn/Al Galvanized Steels Using the Scanning Vibrating Electrode Technique (SVET), Electrochem. Soc. Transactions 13, 2008, pp. 95 ff

[61] Grundmeier, G., Stratmann, M. [9], pp. 590 ff

[62] Grundmeier, G., Schmidt, W., Stratmann, M., Corrosion Protection by Organic Coatings: Electrochemical Mechanism and Novel Methods of Investigation, Electrochim. Acta 45, 2000, pp. 2515 ff

[63] Grundmeier, G., Wapner, C., Stratmann, M., Applications of a New Height Regulated Scanning Kelvin Probe for the Study of Polymer/Metal Interfaces in Corrosive Environments, ICEPAM Conf. Oslo, SINTEF, Trondheim 2004, www.sintef.no/static/mt/norlight/ICEPAM/09-Grundmeier_Max-Planck.pdf ; 01-08-13; 21:30 h

[64] Klimow et al. [46], p. 1291

8 Research topics

8.1 Introduction

Environmental and health awareness has not bypassed the coil coating industry at all. Being a very efficient means of coating metal, thereby rendering durability to a very versatile, but delicate construction material, making the best use of water, treatment chemicals, paints and energy, and allowing for a low-waste and minimum-emission operation, is not just enough.

The striving for heavy metal and, in particular, chromium-free technology in pretreatments and paints has been mentioned previously in Chapters 4.4. and 5.7. Also novel application and curing technologies have already been referred to in Chapters 5.4 and 6.3.2 for alternative application techniques for liquid and powder coatings, as well as Chapters 5.3 and 6.4 for curing by induction, UV or EB radiation, IR or NIR. The 2-in-1 primer-pretreatment (cf. Chapter 4.5.2) approach is now followed by various parties. It will suit particularly well with compact line [1, 2] (cf. Chapter 6.4.1.2) or in-line coating concepts [3].

Most of the research and development effort in the coil coating industry today is therefore targeted not only on further incremental performance and economy improvements, but on gains in additional functionality of the coatings, contributions to climate and environment saving, and a better utilisation of resources.

The present chapter will highlight some further development lines that are recently followed by the industry as well as by academic and institutional research establishments.

8.2 Recycling and renewable materials

Inevitably, organic coatings, paints and varnishes are made up of organic resins, they contain organic materials like wetting and flow agents, and are, most commonly, dissolved in organic solvents. The majority of these materials are made of precursors extracted from petroleum, coal and natural gas. In order to save these fossil primary materials, an interesting approach is being chosen when carbon monoxide, CO, and carbon dioxide, CO_2, the ultimate end products of the oxidative breakdown of organic matter, are recycled and used in the synthesis of polymers and polymer synthons. The products obtained in this way have been introduced, apart from other uses, in the manufacture of paints and coatings, foils for flexible packaging and solvents. Quite obviously they have a favourable CO_2 footprint, as CO_2 is consumed, not released during their manufacture, and also large savings in energy usage are being claimed [4].

Rather than oleochemicals, also raw materials for resin manufacture are being investigated that can be derived from renewable, biological resources. Vegetable oils used for the manufacture of alkyd paints are just one example. Alkyds are potential replacements for fully synthetic polyesters, as they share the same backbone structure. However, being sourced from natural crop, the consistent quality and composition of vegetable oils is an issue. Refined and semisynthetic modified materials nevertheless are being considered [5]. Fats and oils can also serve as precursor for a variety of materials of several chemical classes. Hydrolysis, transesterification, saponification or aminolysis yield the respective fatty acids and esters, carboxylate soaps or amides, along with the glycerol. The latter, in turn, is the starting point for the manufacture of diols, acrylic acid or epichlorohydrine, all of which are raw materials for the production of resins for coil coatings [6]. By methylation of fatty acids, e.g. of rapeseed, linseed or tall oil, semi-synthetic solvents can be made. In polyester synthesis, they can also act as reactive diluents that are consumed in the polymer generation, while allowing control of the viscosity of the reaction mix. Promising performance data of the finally synthesised polyester have been reported [7]. Raw materials for polyester synthesis can be generated from natural precursors. Adipic acid from the oxidation of castor oil, fermentation products like lactic and the isomeric 3-hydroxy propionic acid, or propane-1,3-diol from glycerol, are but some examples. Aromatic polyols or ethers might be derived by pyrolysis from lignin. Being mindful for the emerging competition for food and raw materials, it is also suggested to exploit waste products like crop shells and stalks as precursors for industrial raw materials [8].

8.3 Functional coatings

A development field of common interest for coil coaters and their suppliers is to furnish coatings with additional functionalities. Table 8.1 shows some examples of product ideas that have recently been pursued in the industry community [9, 10].

Anti-fingerprint coatings are of special interest for teletronics applications, as computer and HiFi equipment housings are usually mounted with a chassis and back plate made of bare galvanised steel (HDG, EG, Galvalume; cf. Chapter 4.2.2). Protective coatings are required to prevent ungainly staining caused through the action of the ingredients of human sweat, like fatty acids, peptides, etc. on the metal. On the other hand, the conductivity of the surfaces is necessary for earthing, which means that any coating must not act as an electrical insulator. Moreover, the coating is required to provide lubrication for metal forming. The complex development of suitable products that are environmentally sound (Cr-free, VOC-free, low-bake, etc.) is ongoing [11].

Climatic efficiency is a particularly compelling topic for coil coating, as the vast majority of precoated metals are directed to the building market. There are two approaches to tackle the task. The defensive position will try to cope with the effects of solarisation by insulation of the metal panels that are exposed to sunlight. Particular programmes have been done related to low energy housing and similar projects. Prepainted metal sheet for roofing and cladding plays a role in this concept. Processed to a **composite material**, e.g. as sandwich equipped with an insulating foam core, panels are mounted on walls, flat or pitched roofs [12]. The visco-elasticity of the polymeric cores or coatings can be exploited for vibrational and sound dampening. Composite materials can also be designed to enhance construction properties like stiffness for light-built structures, flooring and roofing. They are also of particular interest to the automotive industry, as weight savings, structural stiffness and also sound-dampening are an increasingly important set of requirements for modern passenger car construction [13].

The **"cool roof"** concept, a proactive view, considers two alternative ways to cope with the solar radiation. One way is to use materials that have a high solar reflectance, hence absorb less of the energy. This will favour the use of light colours in roofing materials. The other approach is being followed by incorporating pigments that absorb near infrared radia-

Table 8.1: Selected development topics in coil coating

Development topics in the coil coating industry
• Anti-fingerprint coatings
• Solar reflectance coatings
- cool roof
• Photovoltaic coatings
• Chromium-free pretreatment with improved humidity resistance
• Combined universal pretreatment-primer
• Interior coatings for fuel tanks/bio-Diesel
• Anti-scaling/self-cleaning coatings
• Anti-bacterial coatings

tion responsible for the heating effect. Suitable pigmentation will downgrade this radiation shining in, and dissipate it through IR fluorescence at lower frequencies, which results in an unaltered appearance in the visible spectral region, hence a range of dark colours remains available which are common with roofing. Respective paints have been invented and commercialised already. ECCA supports the "cool roof" activities with a participation in the EU Cool Roof Council programme [8, 14-18].

The idea of organic **photovoltaic**, OPV, coatings is to replace current vitreous solar cells that have proven to be too expensive to compete in the global market. It is expected that OPV will be cost efficient and easily applicable. By the end of the decade, prognosis forecasts a common use in building and automotive applications [19].

Metal oxides, in particular TiO_2, have been known for **photocatalytic** effects that can be exploited for rendering glass surfaces mist-free, providing anti-bacterial, or anti-soiling properties or even serving for water and air purification [20, 21]. The latter may include destruction of odours as well as the abatement of pollutant toxic gases like NO_x and SO_x. NO_x is generated by combustion processes in car engines, and by several other high-temperature processes, e.g. glass and cement manufacture [22]. Despite the use of catalytic exhaust converters in cars, the 2012 levels of NO_x in the atmosphere, especially in urban areas, exceeded the accepted limits in seven EU member states, incl. Belgium, France and Germany. Coil coated surfaces furnished with active TiO_2 pigmentation have been recently commercialised that are claimed to efficiently provide surfaces cleaning the air in their vicinity. Compatibility with organic paints obviously is critical, as the photocatalytic mechanism involves the generation of hydroxyl radicals that are likely to destroy the organic binder itself. The rollout is currently ongoing [23-28].

The repulsion of soil and scale to surfaces is addressed by designing the **surface energy**. Soiling is, of course, not restricted to metal, but soil will also build up on painted surfaces. The assessment of the resistance of a prepainted metal surface to soiling (dirt pick-up and striping) is subject of a related standard, EN 13523-29. One way of creating a surface with a controlled hydrophilicity is its coating with a layer that is suitably equipped. It has been shown that hydrophobic surfaces generally are less prone to the adsorption of substances. For example, polysiloxane layers with suitable functionalisation (e.g. with alkyl or fluoro alkyl silane terminals), can be obtained by so-called sol-gel processes (cf. Chapter 4.5.1). Similar functionalisation can be considered for topcoats or protective clearcoats by incorporation of monomers with suitable functional terminal groups. However, the situation is not that simple, as the adsorption of different ionic species, for example, may depend on the size (charge density) of the ion, and whether a cation or anion is considered [29].

8.4 Nanotechnology

Like in many other fields and applications of research, nanotechnology is being studied in the development of conversion coatings and paints for coil coating [30]. Nanoparticles are suggested for several purposes, like pigmentation, as carriers for anticorrosive substances, for surface energy control, as reinforcements for structural and anti-scratch properties, as anti-bacterial or anti-fouling agents etc. Nanoparticulate oxides have been used for some time already as additives to improve scratch and mar resistance, UV resistance or enhance photocatalytic activity of clearcoats [31]. Zinc molybdophosphate and aluminium polyphosphate commercial anticorrosive pigments, when milled down to particle sizes between 100 and 200 nm, show improved barrier properties and corrosion resistance in an epoxy primer formulation [32]. Hydrophobically treated silica nanoparticles improved elastic properties (modulus, tensile strength, elongation) of silicone elastomers. It was also shown that, due to the surface energy and hence the adsorption of biological scale being substantially decreased, the elastomer was rendered resistant to fouling [33]. Blends of spherical zinc dust and colloidal lithium silicate, both micronized and as nanoscale preparations, show improved corrosion resistance in NaCl immersion and salt spray tests, the higher the fraction of nanoscale material [34]. Consequently, nanotechnology also plays an important role in scientific research (cf. following Chapter 8.5).

8.5 Academic and institutional research lines

Several strategies are followed in current academic and institutional R&D to develop novel anticorrosive coatings. For coil coating, the following approaches appear interesting.

Polymers like polyaniline, polypyrrol or polythiophene (intrinsically **conductive polymers**, ICP) are being used as pigments in paint compositions for heavy-duty engineering like bridge or heavy machinery construction. Their effect is explained by the redox features of the polymers themselves that can reversibly interfere with the corrosive system and keep the base metal intact [35]. The preparation of thin coatings, however, has not yet emerged from the laboratory stage. Moreover, direct galvanic contact between the polymers and the base metal, as is very likely in thin coatings, forms a high risk to drive the corrosion instead of preventing it. It has been argued that the nature of red-ox active ICP will promote, rather than inhibit, corrosion on large failure sites (scratches, stone-chips, etc.). Nonetheless, an approach to polymerise the ICP in situ in presence of oxoanions, e.g. nitrate, molybdate, or vanadate that may passivate metals, will lead to the formation of a polymer film that hosts to the anions. Rather than continuously leaching out of the coating (like most active anticorrosive pigments do), anions from an ICP film might be released upon changing pH or ion concentrations in the surrounding electrolyte, i.e. under the conditions of corrosion onset, becoming available for the passivation of freshly exposed metal after a damage. The speed of release appears to be the trigger for success [36, 37].

Multilayered polyelectrolytes are known to form covers and membranes. Their use for the functionalisation of textiles and textile fibres is close to commercialisation. On metals, useful barrier properties are observed. However, as the formation of such layers involves repeated iterative treatment steps (layer-by-layer, LbL technique) with polycationic and polyanionic species, the approach is far from being feasible in coil coating [38].

Biopolymers like polysaccharides (e.g. starch, cellulose and derivatives, chitosan) are being investigated because of their natural occurrence. They might be used as polyelectrolytes, or in composites, e.g. with hydroxyl apatite or polysiloxanes [39].

The self-healing capability of an organic coating is one of the topics thoroughly studied. The modern concept involves nano-traps or nano-containers whose voids are filled with inorganic film formers like metal ions that form dense oxide film barriers, polymerisable monomers, inhibitors or hydrophobicisers. Release of these active substances upon particular influences (mechanical damage, electrolyte changes at the onset of a corrosive event, etc.) would then enable active defect healing in the coating [40].

Table 8.2: New concepts in corrosion protection

Nanocontainers	
• Mesoporous SiO_2 and TiO_2 particles with polyelectrolyte (polymer) modified shell • Halloysites (native alumosilicates) with modified shell • Polymer containers with oil containing core and sensitive shell • Polyelectrolyte multilayers and capsules	
Surfaces	
• Metals, polymers, glass	
Encapsulated active substances	
• Various inhibitors (benzotriazole, 8-hydroxy quinoline, molybdates, polyphosphoric acid, ~ esters) • Water repellants (organo-silanes) • Sealants (cyanoacrylates, polyurethanes) • Bio-active substances (vitamins, pharmaceuticals)	
Employed trigger functions	**Layer matrices**
• pH value changes • Mechanical damage • UV radiation • IR radiation	• Sol-gel coatings • Epoxy coatings • Polyacrylates
Technical problems	
• Large-scale container production • Incorporation of loaded containers into the respective coating (agglomeration issues!)	

Nanoscale materials and composites have shown interesting potential to improve anticorrosive properties, and hence have been intensively studied since more than a decade. Particularly, composite materials have attracted interest which might combine functions like anticorrosive properties, self-cleaning and antibacterial features, barrier effects, and controlled release [41, 42]. The loading of a coating with **nanoparticulate carriers** offers another approach to active corrosion protection and self-healing. The preparation and uses of such carriers are subject of funded research activities. Table 8.2 summarises current approaches in this R&D sector. Carriers can be made from LbL-prepared nanosized containers or from synthetic or mineral, porous particles. They can be impregnated with active species like corrosion inhibitors that are slowly released. There are mechanisms known that would allow the controlled release of the active species in the event of a starting corrosion process. Other optional mechanisms include ion exchange or ion scavenger functions. Preparations of nanocontainers are made on the laboratory scale, however obviously, any industrial process is far from realisation [43, 44].

8.6 Literature

[1] Sander, J., Novel Surface Treatment of Metal Strip in the Coil Coating Process, Millennium Steel 2009, pp. 146 ff

[2] Bielefeld, F. W., Sander, J., The Profitable Choice: From small-batch post-painting to Compact Coil Coating, European Coatings Market Day Coil and Can Coatings, Proc., Vincentz Network, Hannover 2011

[3] Jandel, L. T., Coating in galvanising lines, ECCA Autumn Congress, Proc., European Coil Coating Association, Brussels 2011

[4] Michel, A., Besamusca, J., Comparison of the Sustainability of Different Coating Resin Technologies, ECCA General Meeting, European Coil Coating Association, Brussels 2011

[5] Manea, M., High solid binders, Vincentz Network, Hannover 2008, pp. 63 ff

[6] Metzger, J. O., Biermann, U., Fette und Öle als nachwachsende Rohstoffe, auch in der Lackchemie (Fats and oils as renewable raw materials …), 77. GDCh Lacktagung, Proc., Ges. Dtsch. Chem. (German Chem. Soc.), Frankfurt a.M. 2012

[7] Sundell, P.-E., Biomass based reactive diluents for thermal cure coatings, ECCA Autumn Congress, Proc., European Coil Coating Association, Brussels 2011

[8] Lowe, C., You can have any colour you like so long as it is green, ECCA Autumn Congress, Proc., European Coil Coating Association, Brussels 2010

[9] Braghi, E., How Novelis has faced and emerged from the economic recession, ECCA General Meeting, Proc., European Coil Coating Association, Brussels 2010

[10] Kranig, W., Coil coatings: Toolbox for a crisis, ECCA General Meeting, Proc., European Coil Coating Association, Brussels 2010

[11] Steinbach, J., Ultradünne Korrosionsschutzbeschichtungen für galvanisierte Substrate (Ultra-thin corrosion protection coatings…), 77. GDCh Lacktagung, Proc., Ges. Dtsch. Chem. (German Chem. Soc.), Frankfurt a.M. 2012

[12] Brown, N., Energy Efficient Building Solutions, ECCA General Meeting, Proc., ECCA Brussels 2011

[13] King, J., Prospects for the Automotive Market, ECCA General Meeting, Proc., ECCA Brussels 2010

[14] Synnefa, A., Santamouris, M., European Cool Roofs Project and Council, ECCA Autumn Congress, Proc., European Coil Coating Association, Brussels 2010

[15] Evans, E., BlueScope Steel – Innovation & sustainability through our products, ECCA Autumn Congress, Proc., European Coil Coating Association, Brussels 2010

[16] Svedung, H., Experiences with Prelaq Energy; ECCA Autumn Congress Brussels, Proc., European Coil Coating Association, Brussels 2010

[17] Braghi, E. [9]

[18] Zvonkina, I. J., Nothhelfer-Richter, R., Hilt, M., Evaluierung der Effizienz von IR remittierenden Beschichtungen (Evaluation of the efficiency of IR remittant coatings), 77. GDCh Lacktagung, Proc., Ges. Dtsch. Chem. (German Chem. Soc.), Frankfurt a.M. 2012

[19] Kranig, W. [10]

[20] Hashimoto, K., Irie, H., Fujishima, A., TiO_2 Photocatalysis, A Historical Overview and Future Prospects, Jap. J. Appl. Phys. 44, 2005, pp. 8269 ff

[21] Linsebigler, A. L., Lu, G., Yates, J. T., Photocatalysis on TiO_2 surfaces: Principles, mechanisms, and selected results, Chem. Rev. 95, 1995, pp. 735 ff

[22] anon., Nitrogen oxides (NOx) abatement with hydrogen peroxide, Technical Bulletin, US Peroxyde, Atlanta GA, USA 2013; www.h2o2.com/industrial/applications.aspx?pid=101&name=Nitrogen-Oxides-Abatement; 09-08-13, 22.19 h

[23] anon., Directive 2001/81/EC of the European Parliament and of the Council of 23 October 2001 on national emission ceilings for certain atmospheric pollutants, Official Journal of the European Union L 309, 2001, pp. 22 ff, www.eur-lex.europa.eu/...

[24] anon., NEC Directive status report 2012, Technical Report, EEA, Copenhagen 2013, p. 6; www.eea.europa.eu/publications/nec-directive-status-report-2012...; 12-08-13, 12:41 h

[25] anon., The EcoClean process, Video, Internet release, Alcoa 2013, www.alcoa.com/bcs/aap_eastman/ecoclean/en/media.asp; 13-08-13, 11:14 h

[26] anon., AQS Partners With Alcoa to Reduce Air Pollution, Fabricating & Metalworking, Alliance Communications, Pelham AL, USA 2013; www.fabricatingandmetalworking.com/2011/08/aqs-partners-with-alcoa-to-reduce-air-pollution/; 12-08-13, 12:49 h

[27] Fujii, H., Kameshima, J., Omoshiki, K., Kitazaki, S., Adachi, S. (inv.), Photocatalyst coated body and photocatalyst coating liquid, EP 2599545 A1, 2013, Toto Ltd., Kukuoka, Japan

[28] Gaszner, K., The self-cleaning effect of photocatalytic architectural paints, European Coatings Congress, Proc., Vincentz Network 2013

[29] Bellmann, C., Calvimontes, A., Caspari, A., Estel, K., Mauermann, M., Harenburg, J., Innovative Oberflächenmodifikationen zur Verminderung von Schmutz- und Kalkablagerungen (Innovative surface modifications for the reduction of soil and chalk deposits), 77. GDCh Lacktagung, Proc., Ges. Dtsch. Chem. (German Chem. Soc.), Frankfurt a.M. 2012

[30] Zavattoni, M., Nanotechnologies in coil coating pretreatment, Int'l. Paint & Coating Magazine 19, 2013, pp. 86 ff

[31] Christ, U., Öchsner, W. P., Nothhelfer-Richter, R., Nanolacke erschließen neue Anwendungen (Nano paints open up new applications), Metalloberfläche 63, 2009, pp. 12 ff

[32] Entenmann, M., Greisiger, E., Maurer, R., Schauer, T., Corrosion protection with nanoscale anticorrosive pigments in coatings, Europ. Coatings J. (06) 2011, pp. 29 ff

[33] Raeisi, E., Ebrahimi, M., Kassiriha, S. M., Studying the Effect of Silica Nano Particles on the Mechanical and Surface Properties of Silicone Elastomers as Fouling Release Coatings, 77. GDCh Lacktagung Bremerhaven, Proc., Ges. Dtsch. Chem. (German Chem. Soc.), Frankfurt a.M. 2012

[34] Canosa, G., Alfieri, P. V., Giudice, C. A., Environmentally friendly, nano lithium silicate anticorrosive coatings, Progr. Org. Coatgs. 73, 2012, pp. 178 ff

[35] Wessling, B., Corrosion prevention with an organic metal (polyaniline): Surface ennobling, passivation, corrosion test results, Materials and Corrosion 47, 1996, pp. 439 ff

[36] Shchukin, D. G., Skorb, E. in: Sander, J. et al.[24], pp. 165 ff

[37] Rohwerder, M., Michalik, A., Conducting polymers for corrosion protection: What makes the difference between failure and success?, Electrochim. Acta 53, 2007, pp. 1300 ff

[38] Shchukin, D. G., Skorb, E. in: Sander, J. et al.[24], pp. 163 ff

[39] Shchukin, D. G., Skorb, E. in: Sander, J. et al.[24], pp. 173 ff

[40] Shchukin, D. G., Skorb, E. in: Sander, J. et al.[24], pp. 182 ff

[41] Sauvant-Moynot, V., Gonzalez, S., Kittel, J., Self-healing coatings: An alternative route for anticorrosion protection, Progr. Org. Coatgs., 63, 2008, pp. 307 ff

[42] Shchukin, D. G., Skorb, E. in: Sander, J. et al., [24] pp. 177 ff

[43] Saji, V. S., Thomas, J., Nanomaterials for corrosion control, Current Science 92, 2007, pp. 51 ff

[44] Shchukin, D. G., Skorb, E. in: Sander, J. et al.[24]

[45] anon., Bayer MaterialScience beendet Projekt mit Kohlenstoff-Nanoröhrchen, (Bayer Material Science terminates project with carbon nanotubes) Farbe und Lack Newsletter 05/13, Vincentz Network, Hannover 2013

9 Can coating

9.1 Introduction: Precoated metal for packaging

Though food packaging materials, as mentioned earlier, are not included in the ECCA statistics of precoated metal, they shall be highlighted in this chapter. A lot of packaging semis are precoated as coils or sheet, employing technology that is very much alike the coil coating process. Semi-finished stock is used for the manufacture of food and beverage cans, can lids and tabs, closures etc. (cf. Chapter 1.6.5, Figure 1.21) [1].

9.2 Substrates and market

Steel substrates for precoated can, can-end, or tabstock include thin-gauge sheet like double cold-reduced carbon steel (DCR), electrolytically tin or chromium plated steel (tinplate, ECCS), or electro-galvanised stock with gauges between 0.15 and 0.5 mm. 2.7 Mt of these so-called tinmill products were produced in Europe in 2011. Though this figure has been halved since 2007 [2], it still represents a total surface of almost 3 billion m^2 to be coated.

Aluminium for the same end uses mostly comprises AA 3000 and 5000 alloys. In 2011, 16 % of all Al rolled products consumed in Europe (EU 27) were dedicated to rigid packaging, which is equivalent to approx. 720 kt [3], or about 2.5 billion m^2 of coated surface. The beverage can market in 2012 made up 59 billion cans for beer and soft drinks in Europe [4]. Foil stock for semi-rigid packaging, e.g. food and petfood containers, is made from AA 1000 in gauges between 50 and 200 µm. Even thinner material is used for bag laminates, blister packs, etc.

9.3 Pretreatment and base coating

While Al foil is usually left without a separate cleaning treatment after annealing, other substrate material is chemically cleaned following the alkaline or acidic processes that have been described earlier (cf. Chapter 3).

Aluminium coil then receives a phosphochromate (green chromate) or Cr-free pretreatment in immersion, spray or no-rinse application (cf. Chapter 4), and a sprayed or, more commonly, roll-coated base lacquer. The base lacquers are mostly epoxy types, due to their good adhesion properties and chemical resistance.

Tinplate and ECCS are usually left without pretreatment after the cleaning.

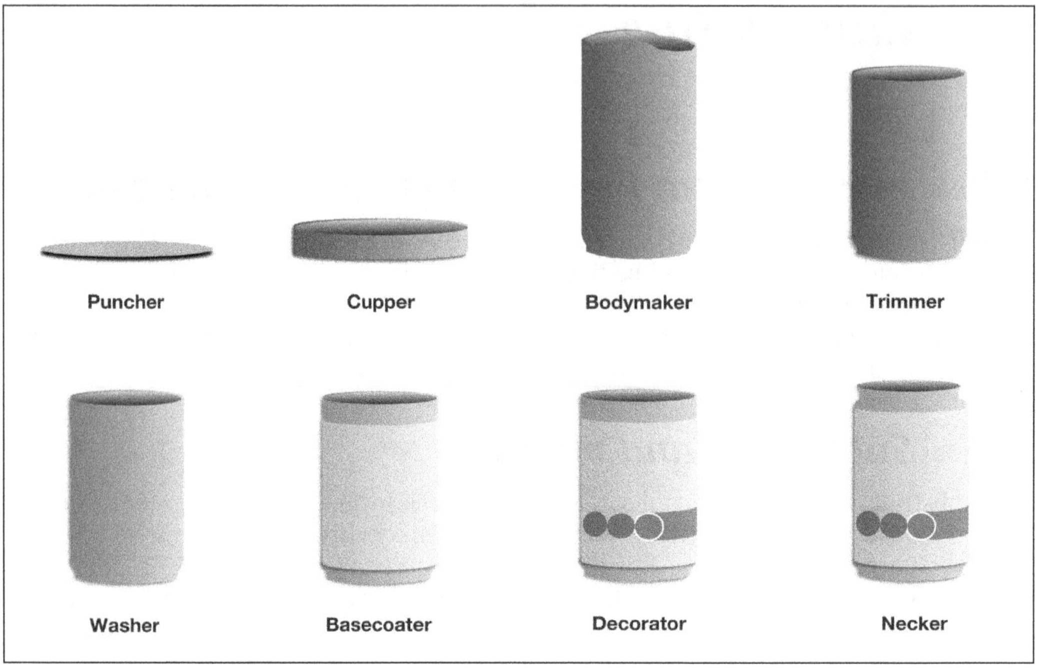

Figure 9.1: Process steps in the manufacture of a beverage can; Al strip that has been prelubricated in a coil line is a familiar feedstock

Coil stock may also be laminated with PET foil etc., either by hot laminating or with an adhesive applied on either foil or metal strip. Al foil, when designated for decorated tops, e.g. for food and petfood containers or blister packaging, is often coated with a hot-sealing adhesive. Al feedstock for beverage cans might be precoated with a dry-film lubricant (prelube). Figure 9.1 illustrates the manufacturing stages of a beverage can starting from a round sheet punched out of Al coil or sheet. Figure 9.2 depicts the pretreatment cycle the bare cans undergo in the washer stage. The cans are processed through the washer line standing upside down on a conveyor belt. They are subjected to a 2 or 3-stage acidic spray cleaning and a Cr-free spray pretreatment, both active stages followed by a rinse cascade. Usually, a mobility enhancer chemical is added to the last DI water rinse to reduce the friction of the can surfaces. In-between stages, excess liquid is blown off from the upward facing can bottoms. Automatic bath control and feeding is common, to allow the process to be operated within narrow limits.

9.4 Can coatings

Most current can coatings are epoxy based, grace to the high barrier and adherence properties of epoxy paints, and their capability for high processing speed. Though the lifetime expectance of a lining for cans, flexible tubes or other food and beverage, or aerosol con-

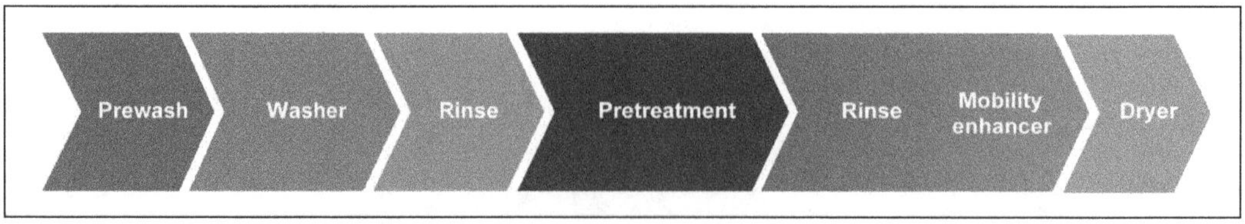

Figure 9.2: Pretreatment cycle in the washer stage of beverage can line

tainers is much shorter than that of e.g. an architectural coil coating film, performance requirements for packaging paints are very high. In particular, extremely low porosity, high resistance against film degradation and leaching of components into the packed goods are mandatory, even under conditions of the canning process which may involve heating and retorting (pasteurisation).

A preferred precursor material of the epoxy paints has been bisphenol A (BPA), and its glycidylated derivative, BADGE. Curing is achieved with co-resins of the phenolic or amino type. Instead of the conventional solventborne paints, high solids and more and more waterborne coatings have taken ground [5].

More recently, epoxy paints are under investigation about apparent adverse effects in the hormonal and endocrine metabolisms of humans, and suspected mutagenic potential. The effects are being disputed in the industry, but nonetheless first regional market bans have taken effect, in particular with respect to baby food [6, 7]. R&D work is therefore ongoing for replacements by polyester based paints. While for Al substrate, these developments are well advanced, an adaptation for other metals is only expected medium term.

9.5 Coil and sheet lines

Can coatings are applied on coil and sheet in installations using coil coating technology or in dedicated sheet coating lines. Large capacity lines are capable of producing well over 50 kt/a. Given the lightweight metal and thin gauge feedstock, down to 150 µm, this corresponds to line speeds of 250 m/min and beyond. Modern line equipment enables swift production with less downtime for coil changes. This includes swivelling edge trimmers that allow width variations between coils under continuous operation, but also automatic adjustment of coater rolls and precise control of the applicator roll pressure that even allows operation versus the floating strip instead. For instance, this enables removal and cleaning of the support roll during coil width changes, avoiding undesired contamination of the reverse side (cf. Chapter 6.3) [8].

Figure 9.3: a) Multiple recoiling unit for aluminium coil at the exit of a strip slitting installation; b) lamination stage for aluminium packaging coil Source: ThyssenKrupp Rasselstein

Figure 9.3a and b show a multiple recoiling unit after strip slitting, and a coil lamination stage for both faces of the coil. Figure 9.4 shows a sheet installation. Single

Figure 9.4: Installation for the coating of aluminium packaging sheet Source: ThyssenKrupp Rasselstein

sheets are fed into this line that can apply single or multiple paint coats for priming, base and varnish coating. The paint is roll-coater applied, and typical curing conditions are 10 min and up to 210 °C.

9.6 Specified tests

Tests particularly dedicated for packaging semis comprise the bend-and-impact test, and deep drawing of cylindrical or quadrangular cups, and simulated food cans etc. Chemical fastness has to be proven in porosity tests, soaking or boiling in organic acids like acetic and lactic acid, and pressurised cooker tests with various media, to assess the sterilisability or retortability. A specialist test for the adhesion of lacquers on can-end stock is the so-called "feathering test". This simulates the opening of a can, by tearing a coated metal specimen apart. Adhesion failures are not accepted which means that no bits of the lacquer must protrude from the torn edge by inspection through the magnifying glass.

The entire coating system, including pretreatment and lacquer, has to pass tests that are designed for the detection of materials migrating from the coating (or the base metal) into the packed goods. The detection is made analytically and by olfactory and flavour tests. The harmlessness of any coating material that is intended for use in direct food contact must be certified by an accredited laboratory.

9.7 Literature

[1] anon., Wege der Produktion - Process Routes, Brochure, ThyssenKrupp Rasselstein, Andernach; www.thyssenkrupp-rasselstein.com/fileadmin/pdf/publikationen/Wege_der_Produktion_DE-EN_Process_Routes.pdf (19-07-13; 15:09 h)

[2] anon., Steel Statistical Yearbook 2012, Worldsteel Assoc., Brussels 2012, p. 44; www.worldsteel.org/dms/internetDocumentList/bookshop/Steel-Statistical-Yearbook-2012/document/Steel%20Statistical%20Yearbook%202012.pdf (19-07-12; 11:01 h)

[3] anon., Aluminium Facts and Figures, European Aluminium Association (EAA), Brussels 2013; www.alueurope.eu/about-aluminium/facts-and-figures/ (19-07-13: 14:13 h)

[4] anon., 2012 Markets - Europe reports, Beverage Can Makers Europe (BCME), Brussels 2013; www.bcme.org/PDF_12/bcme_europe.pdf (19-07-13: 14:46 h)

[5] anon., The short and the long story of epoxy resins, Royal Dutch Shell (ed.), Amsterdam 1992, pp. 39 f

[6] anon., Bisphenol A (BPA) Fact Sheet FCC 461/11, Food Contact Commission, European Metal Packaging (EMPAC), Brussels 2012; www.empac.eu/uploads/downloads/PositionPapers/Empac_BPAfactSheet.pdf (19-07-13; 15:19 h)

[7] anon., EPA Moves Step Closer to Possible Regulatory Action on Bisphenol A, J.Architect. Ctgs., Technology Publishing Co., Pittsburgh, PA, USA 2011

[8] Rieth, B., Alcoa Europe - Successful modernisation of Russian rolling plants, Aluminium 85, 2009, pp. 34 ff

10 Standardisation

10.1 Introduction

Like every industrial activity, the coating of metallic work pieces is subject to numerous standards, norms, guidelines and regulations [1]. These have been primarily developed to ensure a common agreement on reliable quality of operational practice and performance. However, gaining importance with the awareness of the protection of resources and welfare, standards are also intended to create a binding framework for sustainable and responsible manufacturing.

Industrial standards are recommendations of commonly accepted definitions and practices with regard to physical features, quality and performance requirements, procedures of operation, manufacture, monitoring and testing, etc. They usually reflect the current best practice as considered among experts in the matter, and therefore must be regularly revised to ensure they always encompass the recent technical and process development. Quite reasonably, the acceptance of industrial standards depends on their relevance and utility for the participants in the respective market.

Standards have no inherent legal status on their own, however can be made binding by laws and directives, or by contractual agreement between parties in the marketplace. In case of a legal dispute over a quality issue, industrial standards are referred to as defining the common understanding of how a process should be run and how a workpiece should be made. Proof of applying an industrial standard is usually considered as first evidence of having done a work correctly.

Table 10.1: Intrinsic value of standards Source: Meuthen, Jandel [1]

Intrinsic value of standards	Quality features of a standardisation based on consensus
Voluntariness	Anti-trust compliance
Wide participation	User acceptance
Consent	Legitimation
Uniformity	Acceptance by SMEs
Consistency	Global availability
Subject relevance	Global acceptance
State of the art	
Economy	
General benefit	
Internationality	

There are a number of institutional bodies that organise the development, compilation, validation and publication of standards and regulations, and administrate their regular revision, as to stay in pace with common industrial practice and scientific and engineering development. Institutions that are involved in standardisation relevant to coil coating comprise, amongst others, the CEN, the German DIN, and ECISS. Since a lot of the relevant standards have to be harmonised worldwide, the ISO and the international branch of the American organisation, ASTM, must also be mentioned.

The lists of selected relevant standards in the following chapters have been taken from a previous compilation [2], updated and amended [3-6].

10.2 Creation of standards by CEN

10.2.1 CEN

As the centralised European standardisation institution, the European Committee for Standardization/Comité Européen de Normalisation, CEN, was founded in 1961 by the EEC and EFTA. Over 30 national member organisations from the European Union and EFTA, 17 affiliate members from East and South-East Europe, the Middle East, and North Africa, 7 associated members (European NGOs representing e.g. the construction, water, and medical industries, trade and consumers, environmental institutions, trade unions, and SMEs), corresponding standardisation institutes from outside Europe, as well as the "Counsellors" (European Commission and EFTA), participate in CEN. CEN also has global cooperation activities to the aim of a most widespread acceptance of industry standards worldwide [1, 7].

10.2.2 Standardisation procedure

10.2.2.1 Proposal stage

Any new European Norm (EN) - if not adapted international standards - is created in CEN's technical committees (TCs) or their sub-committees (SCs), or working groups (WGs). All members are compulsory members to a TC or SC. Active participation in one or more WGs is encouraged. Working out a standard or any other relevant document is carried out in consecutive stages.

Proposals for the creation of a European Standard, a Technical Specification or a Technical Report, including revisions and amendments, can be put in by any member, the technical bodies, the European Commission, the EFTA Secretariate, European or international groups. Based on the information provided by the applicant, the CEN/TC in charge first evaluates the relevance, usefulness, and feasibility of the proposed standardisation project. Approval requires the interest and active cooperation of at least five CEN members, and a realistic prospect to finish work within three years.

10.2.2.2 Working stage

As soon as a standardisation project has entered the working stage, no single CEN member is entitled anymore to issue an individual standard that might interfere (Stand Still Rule). Publications may however be issued to help in the formation of opinion. The working stage comprises, in repetitive work cycles, the creation of a lead document and a draft standard (provisional European Norm, prEN). This provisional standard is then submitted to the next stage.

10.2.2.3 Enquiry stage

Within five months after submission, the draft standard has to be reviewed and commented by all CEN members. Comments and amendment proposals are collected and reviewed by the involved CEN/TC or its authorised working group. At this stage, a standardisation can be processed into a Final Draft European Standard (FprEN) for formal approval, or revised to an alternative draft for further negotiation, or it can be downgraded to a Technical Specification, or abandoned altogether.

10.2.2.4 Approval stage

During the two-month approval stage, editorial corrections of the draft standard can still be made, however otherwise only approval or (reasoned) decline is possible. To become effective, the FprEN must be approved by a classified majority (71 %) of CEN members.

10.2.2.5 Implementation stage

The implementation comprises final editing, translation into English, French and German, and distribution of the final EN standard to the CEN members. Transition into national standards and the withdrawal of conflicting national standards are up to the single members.

Table 10.2: National members of the Comité Européen de Normalisation, CEN Source: CEN

National members of CEN		
Austria	Greece	Norway
Belgium	Hungary	Poland
Bulgaria	Iceland	Portugal
Croatia	Ireland	Romania
Cyprus	Italy	Slovakia
Czech Republic	Latvia	Slovenia
Denmark	Lithuania	Spain
Estonia	Luxembourg	Sweden
Finland	Macedonia	Switzerland
France	Malta	Turkey
Germany	The Netherlands	United Kingdom

Table 10.3: Affiliate members of the Comité Européen de Normalisation, CEN
Affiliate membership in CEN requires membership in or affiliation to ISO Source: CEN

Affiliate members of CEN		
Albania	Georgia	Montenegro
Armenia	Israel	Morocco
Azerbaijan	Jordan	Serbia
Belarus	Lebanon	Tunisia
Bosnia and Hercegovina	Libya	Ukraina
Egypt	Moldova	

Furthermore, any European Standard is subject to a regular Systematic Review to be performed after five years at longest. Whatever appropriate, in this review process the EN can be confirmed, amended, superseded by an ISO Standard that may have issued in the meantime, or withdrawn.

10.3 ECISS, ISO and ASTM

Formally independent, but working in the framework of CEN, the European Committee for Iron and Steel Standardization, **ECISS**, specialises in the standardisation of definitions, classifications, test procedures, chemical analysis and technical specifications of steel products. The organisation superseded the former COCOR institution of the European Community for Coal and Steel in 1986. COCOR had its roots in the first European integration activities in 1952/53. Its standards had no binding effect. Today, the ECISS draft standards, however, are regularly fed into the CEN approval procedures.

The International Organization for Standardization, **ISO**, was founded 1947 as a global federation of non-governmental standardisation bodies. It is headquartered in Geneva, and currently has 112 full, and 47 correspondent members. Four more members have subscribed. This means that, with the exception of some central African states, all

Table 10.4: Important standardisation organisations for coil coating

Standardisation institutes
• CEN, Comité Européen de Normalisation, Brussels; www.cen.eu, www.cenorm.be
• DIN, Deutsches Institut für Normung e.V., Berlin; www.din.de
• ECISS, European Committee for Iron and Steel Standardization, Bussels, CEN Associated Standards Body; www via CEN
• ISO, Int. Organization for Standardization, Geneva; www.iso.org
• ASTM International, American Society for Testing and Materials, W. Conshohocken (PA), USA, www.astm.org

countries of the world are represented in ISO. Correspondent and subscriber members have restricted access to the standardisation process as observers, but cannot influence it.

ISO Standards are created in a similar procedure as for the European Standards. Official versions are published in English and French. They are not compulsory, unless, for CEN members, they have been adopted as an official European Standard by CEN. To-date, ISO has published more than 19,500 international standards during its existence [8].

The American Society for Testing and Materials, ASTM, was founded in 1883. Through its worldwide branch, **ASTM International**, its activities overlap with those of ISO. About 12,000 ASTM standards concerning quality, safety, industrial practice and testing, have issued since the beginnings. They are globally accepted, and applied by convention between industrial parties on a voluntary basis [9].

10.4 General standards and regulations

10.4.1 Relevant standardisation bodies

CEN – European Committee for Standardization/Europäisches Komitee für Normung/Comité Européen de Normalisation, Rue de Stassart, 36, B-1050 Brussels; www.cen.eu, www.cenorm.be

CEN/TC 132: Aluminium and aluminium alloys; Secretariate: AFNOR, Paris

CEN/TC 132/WG 7: Sheets, strips and plates

Normenausschuss Nichteisenmetalle (FNNE) im DIN Deutsches Institut für Normung e.V. Burggrafenstr. 6, D-10787 Berlin, www.fnne.din.de (Standardisation Board Non-Ferrous Metals in the German Standardisation Institute, DIN)

ECISS – European Committee for Iron and Steel Standardization (CEN Associated Standards Body)/Europäisches Komitee für die Eisen- und Stahlnormung/Comité Européen de Normalisation du Fer et de l'Acier, Rue de Stassart, 36, B-1050 Brussels

ECISS/TC 27: Surface coated flat products: Qualities, dimensions, tolerances and specific tests, NA 021-00-01-02 UA, Standardisation Board Iron and Steel in the German Standardisation Institute, DIN

ECISS/TC 27/SC 1: Continuously organic coated (coil coated) steel flat products, NA 021-00-01-03 UA Standardisation Board Iron and Steel in the German Standardisation Institute, DIN

Normenausschuss Eisen und Stahl (FES), in Standardisation Board Iron and Steel in the German Standardisation Institute, DIN, Sohnstr. 65, D-40237 Düsseldorf, POB 105145, D-10042 Düsseldorf

10.4.2 *Coil coated metal: Terminology and definitions*

Terms and definitions as they are commonly used in the industry and also reflected in the relevant standards are listed hereunder in alphabetical order. Definitions may vary slightly between different standards.

Accessibility: Ease of access to a coated building component for inspection and maintenance

Aluminium, Al: Metal with a minimum Al content of 99.0 %, providing the mass fractions of other elements are

- Iron (Fe) and silicon (Si): < 1.0 %
- Copper (Cu): < 0.2 %, providing that each of chromium (Cr) and manganese (Mn) are present with < 0.05 %
- All other elements: < 0.1 %

Aluminium alloy: Metallic substance whose largest constituent is Al, and where the mass fractions of other elements comply with individually specified ranges outside the above definition

Backing coat: Any single-coat protective layer on the reverse side of a coil, without specific requirements

Basic material, substrate: Basic product, manufactured from aluminium or its alloys by cold rolling

Building exterior application: Any architectural application that renders components exposed to the atmosphere. NOTE: Components are e.g. profiled sheets for roof and wall, extruded or roll-formed profiles, standing seam, corrugated or tile-shaped sheet, hidden fastening components for roof and wall, sandwich components, rainware (gutters and drains), metal doors and garage doors. Corrosive load must be considered also for the reverse side, in particular under conditions of high humidity or chemical pollution inside the building

Building interior application: All architectural uses where components are used in the building interior (incl. e.g. structural building components), and neither face of the coil coated material is exposed to the outside atmosphere. NOTE: In case one face of the coil coated component is exposed to the atmosphere, EN 10169-2 (Building external applications) applies. Components are e.g. partition walls, flooring and ceiling parts (frames, panels), sandwich components for cool storage rooms or rooms with controlled environment, door frames, metal doors and window frames for indoor use. For lighting components, special requirements may apply

Coating ductility class: Classification of a continuously coil coated flat product concerning its strain resistance

Coating material: Material suitable for coil coating that contains organic polymers, e.g. resins or plastics, and usually pigments, auxiliary additives and, if applicable, solvents. Liquid and powder paints or plastic foils are possible

Coating system: Combination of coatings that are applied on either side of the substrate. The material of the top coating is relevant for the designation of the total coating system *(EN 1396: 2007)*. Aggregate of all coatings applied on either top or reverse side of a coil, consists of one or more layers of one or more coating materials. The designation of a coating system is derived from its outermost (top) layer *(EN 10169-1: 2004)*

Coating thickness, coating gauge: Total thickness of a coating on each side of the coil

Coil coating: Continuous process for the application of a coating material on to a cold rolled metal coil, comprising cleaning and a chemical pretreatment of the surface, and either

- application of a liquid or powder paint (single or double sided, each in one or more layers) which is subsequently oven cured, or
- lamination with a plastic foil

Colour/colour difference or distance: The visual perception of electromagnetic radiation of a particular spectral composition creates the sensation of colour. A colour can be distinctly characterised as a pair of coordinates in the colour space (colour metrics designation by standard colours, under standard conditions). The difference in the size (brightness) and relative spectral composition (hue) defines the colour distance of two colours (corresponding to EN 13523-3)

Corrosion load: Environment conditions that further corrosion (cf. EN ISO 12944-2)

Corrosion protection class: Classification of the corrosion protection properties of a coating, considering the corrosivity, the service term and the accessibility of the coated object

Corrosion system: System composed of one or more metals and all environmental conditions that affect corrosion (cf. EN ISO 8044)

Corrosivity: Capability of a medium to cause corrosion in a preset corrosion system (cf. EN ISO 8044)

Corrosivity class: Category describing the corrosive effect of an environment, concerning local and micro climates

Duration of moistening: Length of coverage of a metallic surface by a liquid electrolyte that is capable of causing atmospheric corrosion (cf. EN ISO 12944-2). NOTE: As approximation, the time total can be calculated during which the temperature is above freezing point and, simultaneously, the relative humidity above 80 %

Environment, surrounding: Conditions prevailing inside a building, determining the local corrosivity classification. NOTE: Conditions comprise e.g. air temperature, relative humidity, workroom conditions, pollution by aggressive chemicals, cooled areas

Film coating, foil coating: Foil (organic film) that is applied on to the substrate that usually has been furnished with an adhesive or, if appropriate, a primer before

Gloss/reflectance: Gloss is brought about by the capability of a surface to reflect light. Reflectance values are defined by the ratio of the light flux (directed reflection) from a specimen versus the light flux from a polished black glass plate (corresp. to EN 13523-3) NOTE: For convenience, single gloss ranges are usually referred to as matte, low-gloss (or semi-gloss), silk-matte, glossy or high-gloss

Intermediate coating: Any coating between a primer and a topcoat (finish coat)

Local environment conditions: Atmospheric conditions around a building (corrosivity classification corresponding to EN ISO 12944-2 and -21). NOTE: The corrosivity classification comprises meteorological factors and pollution. Local conditions might be substantially different from the regional situation

Master coil, parent coil: Coil that has been coated as a unit and serves as feedstock for cut-to-size, tailored products (coil, slit strip, or sheet)

Micro climate: Conditions at the interface between a component and its environment (cf. EN ISO 12944-2), most influential on the actual corrosion load

Multiple-coat system: Consisting of a primer or precoat, if applicable, one or more intermediate coatings and a finish coat. It matches particular requirements regarding the surface aspect (decorative effect), the formability, the corrosion protection, the subsequent coloured coating, etc.

Nominal coating thickness: Coating thickness according to the order or specification NOTE: The nominal thickness of the coated product is equivalent to the substrate gauge!

Operational monitoring: Regular internal supervision of conditions

Organic coating: Dry paint film of the coated material, or organic film of the film-metal laminate

Original inspection: First performed tests to prove the accordance of a product with the applicable standard

Performance test: Quality assessment shall consider realistic utilisation conditions of a material

Protection period: Term between the start date of a component's service and the instant of first maintenance to recover corrosion protection. NOTE: Maintenance is usually considered necessary when a certain fraction of the component's coating fails enough to allow structural corrosion of the substrate

Reverse side: Inner face of a coil, usually coated with a backing coat, or, if applicable, a specialist coating that provides particular features, e.g. adhesion to construction foams (polyurethane)

Saturated colour: Colour of an intensity (chroma value) $C > 45$

Single-coat system: Coating consisting of a single layer, matching particular requirements regarding the surface aspect (decorative effect), the formability, the corrosion protection, the subsequent coloured coating, etc. When applied as primer, the coating provides specific features concerning e.g. adhesive bonding or corrosion protection for the finish coating that is applied during a later, batch coating process

Strippable foil: Plastic foil applied on to a coated surface to provide temporary protection against mechanical damage

Top side: Coil side with the higher quality (decorative surface aspect), facing upwards in normal operations. Usually, the topside is the outer face of a coil, or the upper face of cut-to-size sheets in a stack

Topcoat, finish coat: Last (top) layer of a multiple-coat system

Uncoated: The condition of a base material surface that has partially, e.g. on one coil side, been left without a coating

UV resistance class: Classification of the UV resistance properties of a coating, considering the UV irradiation (class), the service term and the aspect requirements of the coated object

10.4.3 Coil coated metal: Product standards

Coil coated products have been standardised under CEN norms EN 1396 (aluminium) and EN 10169 (steel). The latter norm has only recently superseded and combined the previous set of standards issued by ECISS. The standards give a general description of the technical specifications and uses.

EN 1396: 2007, Aluminium and aluminium alloys - Coil coated sheet and strip for general applications - Specifications

EN 10169: 2010+A1: 2012, Continuously organic coated (coil coated) steel flat products - Technical delivery conditions

10.5 Substrate and test standards

10.5.1 Overview

Substrates relevant for coil coating are steel, in particular galvanised steel varieties, and several wrought aluminium alloys. All of them are mentioned in the umbrella norms for coil coating, the EN 1396 and 10169, while the material specifications may refer to other standards, like the Aluminium Association's nomenclature system of alloys. The new ternary-alloyed galvanised steel substrate, which contains Al and Mg, has only recently been standardised in a provisional material sheet issued by the German Steel Institute.

The compilation is completed by standards concerning less important substrates for coil coating, like hot-rolled carbon steel, electrical steel and stainless steel. Finally, also substrates are covered that are commonly found in packaging applications and can coating.

10.5.2 Coil coated aluminium

Standardisation Body: NA 066-01-06 AA

EN 485-1: 2008+A1: 2009, Aluminium and aluminium alloys - Sheet, strip and plate - Part 1: Technical conditions for inspection and delivery

EN 485-2: 2008, Aluminium and aluminium alloys - Sheet, strip and plate - Part 2: Mechanical properties

EN 485-4: 1993, Aluminium and aluminium alloys - Sheet, strip and plate - Part 4: Tolerances on shape and dimensions for cold-rolled products

EN 515: 1993, Aluminium and aluminium alloys - Wrought products - Temper designations

EN 541: 2006, Aluminium and aluminium alloys - Rolled products for cans, closures and lids - Specifications

EN 573-1: 2004, Aluminium and aluminium alloys - Chemical composition and form of wrought products - Part l: Numerical designation system

EN 573-2: 1994, Aluminium and aluminium alloys - Part 2: Chemical symbol based designation system

EN 573-3: 2009, Aluminium and aluminium alloys - Chemical composition and form of wrought products - Part 3: Chemical composition and form of products

EN 573-5: 2007, Aluminium and aluminium alloys - Chemical composition and form of wrought products - Part 5: Codification of standardized wrought products

EN 602: 2004, Aluminium and aluminium alloys - Wrought products - Chemical composition of semi-finished products used for the fabrication of articles for use in contact with foodstuff

EN 683-1: 2006, Aluminium and aluminium alloys - Finstock -
Part 1: Technical conditions for inspection and delivery

EN 683-2: 2006, Aluminium and aluminium alloys - Finstock -
Part 2: Mechanical properties

EN 683-3: 2006, Aluminium and aluminium alloys - Finstock -
Part 3: Tolerances on dimensions and form

EN 15530: 2008, Aluminium and aluminium alloys - Environmental aspects of aluminium products - General guidelines for their inclusion in standards

10.5.3 Coil coated steel

10.5.3.1 General provisions

Standardisation Body: NA 021-00-01-03 UA (ECISS/TC 6, TC 7, TC 9)

EN 606: 2004, Bar coding - Transport and handling labels for steel products

EN 10020: 2000, Definition and classification of grades of steel

EN 10021: 2006, General technical delivery conditions for steel products

EN 10027-1: 2005, Designation systems for steels - Part 1: Steel names

EN 10027-2: 1992, Designation systems for steels - Part 2: Numerical systems

EN 10079: 2007, Definition of steel products

EN 10168: 2004, Steel products - Inspection documents - List of information and description

EN 10204: 2004, Metallic products - Types of inspection documents

CR 10313: 2000, Classification of grades of steel - Examples of classification rerated to European Standards

10.5.3.2 Cold-rolled steel substrates

Standardisation Body: NA 021-00-01-UA (Delivery Conditions) and NA 021-00-20-02 UA (Tolerances) (ECISS/TC 13)

EN 10130: 2006, Cold rolled low carbon steel flat products for cold forming -
Technical delivery conditions

EN 10131: 2006, Cold rolled uncoated and zinc or zinc-nickel electrolytically coated low carbon and high yield strength steel flat products for cold forming - Tolerances on dimensions and shape

EN 10139: 1997, Cold rolled uncoated mild steel narrow strip for cold forming -
Technical delivery conditions

EN 10140: 2006, Cold rolled steel narrow strip - Tolerances on dimensions and shape

EN 10268: 2006, Cold rolled steel flat products with high yield strength for cold forming - Technical delivery conditions

prEN 10338: (under approval: 2016), Hot and cold rolled non-coated flat products of multiphase steels for cold forming - Technical delivery conditions

10.5.3.3 Metallic-coated steel substrates (except packaging sheet)

Standardisation body: NA 021-00-01-02 UA (Delivery Conditions and Tolerances) (ECISS/TC 27)

Stahl - Eisen Werkstoffblatt (Steel - Iron Construction Material Sheet) SEW 022, 1st ed.: Continuously hot-dip coated steel flat products - Zinc-magnesium coatings, technical delivery conditions, German Steel Institute, Düsseldorf 2012 [10]

EN 10143: 2006, Continuously hot-dip coated steel sheet and strip - Tolerances on dimensions and shape

prEN 10152: 2009/AC: 2011, Electrolytically zinc coated cold rolled steel flat products for cold forming - Technical delivery conditions

EN 10271: 1998, Electrolytically zinc-nickel (ZN) coated steel flat products - Technical delivery conditions

prEN 10346: (under approval: 2015), Continuously hot-dip coated steel flat products - Technical delivery conditions

10.5.3.4 Further cold rolled and metallic coated steel substrates - Packaging sheet

Standardisation body: NA 021-00-02 AA (ECISS/TC 26)

EN 10202/AC: 2003, Cold reduced tin mill products - Electrolytic tinplate and electrolytic chromium/chromium oxide coated steel

EN 10205: 1991, Cold reduced black plate in coil form for the production of tinplate or electrolytic chromium/chromium oxide coated steel

EN 10333: 2005, Steel for packaging - Flat steel products intended for use in contact with foodstuffs, products and beverages for human and animal consumption - Tin coated steel (tinplate)

EN 10334: 2005 Steel for packaging - Flat steel products intended for use in contact with foodstuffs, products and beverages for human and animal consumption - Non-coated steel (black plate)

EN 10335: 2005, Steel for packaging - Flat steel products intended for use in contact with foodstuffs, products and beverages for human and animal consumption - Non alloyed electrolytic chromium/chromium oxide coated steel

10.5.3.5 Hot rolled steel substrates

Standardisation body: NA 021-00-20-01 UA (ECISS/TC 13)

EN 10025-1: 2004, Hot rolled products of structural steels - Part 1: General technical delivery conditions

EN 10025-2: 2004/AC: 2005, Hot rolled products of structural steels - Part 2: Technical delivery conditions for non-alloy structural steels

EN 10025-3: 2004, Hot rolled products of structural steels - Part 3: Technical delivery conditions for normalized/normalized rolled weldable fine grain structural steels

EN 10025-4: 2004, Hot rolled products of structural steels - Part 4: Technical delivery conditions for thermomechanical rolled weldable fine grain structural steels

EN 10025-5: 2004, Hot rolled products of structural steels - Part 5: Technical delivery conditions for structural steels with improved atmospheric corrosion resistance

EN 10025-6: 2004+A1: 2009, Hot rolled products of structural steels - Part 6: Technical delivery conditions for flat products of high yield strength structural steels in the quenched and tempered condition

EN 10048: 1996, Hot rolled narrow steel strip - Tolerances on dimensions and shape

EN 10051: 2010, Continuously hot-rolled strip and plate/sheet cut from wide strip of non-alloy and alloy steels - Tolerances on dimensions and shape

prEN 10111: 2008, Continuously hot-rolled low carbon steel sheet and strip for cold forming - Technical delivery conditions

10.5.3.6 Electrical steel

Standardisation Body: NA 021-00-07 GA (GA FES/DKE) (ECISS/TC 24)

EN 10106: 2007, Cold rolled non-oriented electrical steel sheet and strip delivered in the fully processed state

EN 10107: 2005, Grain-oriented electrical steel sheet and strip delivered in the fully processed state

EN 10265: 1995, Magnetic materials - Specification for steel sheet and strip with specified mechanical properties and magnetic permeability

EN 10303: 2001, Thin magnetic steel sheet and strip for use at medium frequencies

EN 10341: 2006, Cold rolled electrical non-alloy and alloy steel sheet and strip delivered in the semi-processed state

EN 10342: 2005, Magnetic materials - Classification of surface insulations of electrical steel sheet, strip and laminations

10.5.3.7 Stainless steels

Standardisation Body: NA 021-00-06-01 UA

EN 10088-1: 2005, Stainless steels - Part 1: List of stainless steels

EN 10088-2: 2005, Stainless steels - Part 2: Technical delivery conditions for sheet/plate and strip of corrosion resisting steels for general purposes

EN 10088-4: 2009, Stainless steels - Part 4: Technical delivery conditions for sheet/plate and strip of corrosion resisting steels for construction purpose

10.5.4 Coil treatment lines: Standards and regulations

EN 1539: 2009, Dryers and ovens, in which flammable substances are released - Safety requirements

EN 12753: 2005+A1: 2010, Thermal cleaning systems for exhaust gas from surface treatment equipment – Safety requirements

EN 12921-1: 2005, Machines for surface cleaning and pre-treatment of industrial items using liquids or vapours – Part 1: Common safety requirements

EN 12921-2: 2005, Machines for surface cleaning and pretreatment of industrial items using liquids or vapours – Part 2: Safety of machines using water based cleaning liquids

EN 14462: 2005, Surface treatment equipment – Noise test code for surface treatment equipment including its ancillary handling equipment – Accuracy grades 2 and 3

EN 15601: 2005, Safety of Machinery – Safety requirements for strip processing line machinery and equipment

EN 50271: 2007+A1: 2008, Electrical apparatus for the detection and measurement of combustible gases, toxic gases or oxygen – Requirements and tests for apparatus using software and/or digital technologies [11]

prEN 50402: 2013, Electrical apparatus for the detection and measurement of flammable gases or vapours or of oxygen - Requirements of the functional safety of fixed gas detection systems (to supersede EN 50402: 2005+A1: 2008) [12]

EN 60079-0: 2012, Explosive atmospheres - Part 0: Equipment - General requirements [13]

Directive 94/9/EC of the European Parliament and the Council of 23 March 1994 on the approximation of the laws of the Member States concerning equipment and protective systems intended for use in potentially explosive atmospheres

Corrigendum to Directive 94/9/EC of the European Parliament and of the Council of 23 March 1994 on the approximation of the laws of the Member States concerning equipment and protective systems intended for use in potentially explosive atmospheres (OJ L 100 of 19.4.1994)

Council Directive 1999/13/EC of 11 March 1999 on the limitation of emissions of volatile organic compounds due to the use of organic solvents in certain activities and installations

Corrigendum to Council Directive 1999/13/EC of 11 March 1999 on the limitation of emissions of volatile organic compounds due to the use of organic solvents in certain activities and installations (OJ L 85 of 29.3.1999) (This corrigendum cancels and replaces the corrigendum published in Official Journal of the European Communities L 188 of 21 July 1999, page 54)

Directive 2002/95/EC of the European Parliament and of the Council of 27 January 2003 on the restriction of the use of certain hazardous substances in electrical and electronic equipment

10.5.5 Test methods

10.5.5.1 Overview

There are a number of generally accepted rules for process control policies in coil coating, although not all of them have already been standardised. It is common practice to monitor the coating process continuously, which means at least that the wet paint film is measured along the entire length of the strip. Often this is done on various positions across the coil width, or the measuring devices are even moved across the strip statistically or in steady oscillation. Methods of online measurement of the coating gauge are standardised under different ISO and EN norms.

Also colour and gloss are monitored online during manufacture, and it is common to observe the curing temperature of the strip surface (peak metal temperature) continuously by IR detection.

The norm EN 13523 forms a series of standards covering performance test procedures and assessment guidelines. Currently 29 parts of this norm have been issued, including coating thickness, colour and gloss, physical features, adhesion, corrosion and chemical resistance. This serial norm has been developed in close cooperation with ECCA's Technical Committee Team 3 that is dedicated to standardisation.

The solid content and yield of a coating material is covered by a new provisional standard, pr EN 16074.

10.5.5.2 EN standardisation body in charge

CEN – European Committee for Standardization/Europäisches Komitee für Normung/ Comité Européen de Normalisation, Rue de Stassart 36, B-1050 Brussels; www cenorm.be

CEN/TC 139: Paints and varnishes

CEN/TC 139/WG 9: Testing of coil coated metals (in cooperation with ECCA/TC 3)

10.5.5.3 EN testing standards

EN 13523-0: 2001, Coil coated metals – Test methods – Part 0: General introduction and list of test methods

EN 13523-1: 2009, Coil coated metals – Test methods – Part 1: Film thickness

EN 13523-2: 2001, Coil coated metals – Test methods – Part 2: Specular gloss

EN 13523-3: 2001, Coil coated metals – Test methods – Part 3: Colour difference – Instrumental comparison

EN 13523-4: 2001, Coil coated metals – Test methods – Part 4: Pencil hardness

EN 13523-5: 2001, Coil coated metals – Test methods – Part 5: Resistance to rapid deformation (impact test)

EN 13523-6: 2002, Coil coated metals – Test methods – Part 6: Adhesion after indentation (cupping test)

EN 13523-7: 2001, Coil coated metals – Test methods – Part 7: Resistance to cracking on bending (T-bend test)

EN 13523-8: 2010, Coil coated metals – Test methods – Part 8: Resistance to salt spray (fog)

EN 13523-9: 2001, Coil coated metals – Test methods – Part 9: Resistance to water immersion

EN 13523-10: 2010, Coil coated metals – Test methods – Part 10: Resistance to fluorescent UV radiation and water condensation

EN 13523-11: 2011, Coil coated metals – Test methods – Part 11: Resistance to solvents (rubbing test)

EN 13523-12: 2004, Coil coated metals – Test methods – Part 12: Resistance to scratching

EN 13523-13: 2001, Coil coated metals – Test methods – Part 13: Resistance to accelerated ageing by the use of heat

EN 13523-14: 2001, Coil coated metals – Test methods – Part 14: Chalking (Helmen method)

EN 13523-15: 2002, Coil coated metals - Test methods - Part 15: Metamerism

EN 13523-16: 2004, Coil coated metals - Test methods - Part 16: Resistance to abrasion

EN 13523-17: 2011, Coil coated metals - Test methods - Part 17: Adhesion of strippable films

EN 13523-18: 2002, Coil coated metals - Test methods - Part 18: Resistance to staining

EN 13523-19: 2011, Coil coated metals - Test methods - Part 19: Panel design and method for atmospheric exposure testing

EN 13523-20: 2011, Coil coated metals - Test methods - Part 20: Foam adhesion

EN 13523-21: 2010, Coil coated metals - Test methods - Part 21: Evaluation of outdoor exposed panels

EN 13523-22: 2010, Coil coated metals - Test methods - Part 22: Colour difference - Visual comparison

EN 13523-23: 2002, Coil coated metals - Test methods - Part 23: Colour stability in humid atmospheres containing sulfur dioxide

EN 13523-24: 2004, Coil coated metals - Test methods - Part 24: Resistance to blocking and pressure marking

EN 13523-25: 2006, Coil coated metals - Test methods - Part 25: Resistance to humidity

EN 13523-26: 2006, Coil coated metals - Test methods - Part 26: Resistance to condensation of water

EN 13523-27: 2009, Coil coated metals - Test methods - Part 27: Resistance to humid poultice (Cataplasm test)

EN 13523-29: 2010, Coil coated metals - Test methods - Part 29: Resistance to environmental soiling (Dirt pick-up and striping)

Various parts of this standard are under revision. A further Draft, Part 28: Evaluation of mildew, has been withdrawn, as its scope is covered by prEN 16492 (see Chapter 10.5.4.4) [14].

10.5.5.4 Test methods: Further standards

(in numerical order)

Standardisation Body CEN/TC 139, ISO/TC 35 and ASTM

ASTM B567: 1998 (2009a), Standard test method for measurement of coating thickness by the beta backscatter method

EN ISO 1518-1: 2011, Paints and varnishes - Determination of scratch resistance - Part 1: Constant-loading method (ISO 1518-1: 2011)

EN ISO 1518-2: 2011, Paints and varnishes - Determination of scratch resistance - Part 2: Variable-loading method (ISO 1518-2:2011)

EN ISO 1519: 2011, Paints and varnishes - Bend test (cylindrical mandrel) (ISO 1519: 2011)

EN ISO 1520: 2006, Paints and varnishes - Cupping test (ISO 1520: 2006)

EN ISO 1522: 2006, Paints and varnishes - Pendulum damping test (ISO 1522: 2006)

ASTM E 1640-09, Standard Test Method for Assignment of the Glass Transition Temperature By Dynamic Mechanical Analysis [15]

SEP 1850: Cyclic corrosion testing of materials and components in automotive construction, Stahl-Eisen Prüfblatt (SEP) 1850 (equivalent to VDA 233-102) [16]

EN ISO 2409: 2013, Paints and varnishes - Cross-cut test (ISO 2409: 2013)

EN ISO 2431: 2011, Paints and varnishes - Determination of flow time by use of flow cups (ISO 2431: 2011)

EN ISO 2808: 2007, Paints and varnishes - Determination of film thickness (ISO 2808: 2007)

EN ISO 2810: 2004, Paints and varnishes - Natural weathering of coatings - Exposure and assessment (ISO 2810: 2004)

EN ISO 2812-1: 2007, Paints and varnishes - Determination of resistance to liquid - Part 1: Immersion in liquids other than water (ISO 2812-1: 2007)

EN ISO 2812-2: 2007, Paints and varnishes - Determination of resistance to liquid - Part 2: Water immersion method (ISO 2812-2: 2007)

EN ISO 2812-3: 2012, Paints and varnishes - Determination of resistance to liquid - Part 3: Method using an absorbent medium (ISO 2812-3: 2012)

EN ISO 2812-4: 2007, Paints and varnishes - Determination of resistance to liquid - Part 4: Spotting methods (ISO 2812-4: 2007)

EN ISO 2812-5: 2007, Paints and varnishes - Determination of resistance to liquid - Part 5: Temperature-gradient oven method (ISO 2812-5: 2007)

EN ISO 2813: 1999, Paints and varnishes - Determination of specular gloss of non-metallic paint films at 20°, 60° and 85° (ISO 2813: 1994, including Technical Corrigendum 1: 1997)

EN ISO 2815: 2003, Paints and varnishes - Buchholz indentation test (ISO 2815: 2003)

EN ISO 3231: 1997, Paints and varnishes - Determination of resistance to humid atmospheres containing sulfur dioxide (ISO 3231: 1993)

EN ISO 3248: 2000, Paints and varnishes - Determination of the effect of heat (ISO 3248: 1998)

EN ISO 3543: 2000/AC: 2006, Metallic and non-metallic coatings - Measurement of thickness - Beta backscatter method (ISO 3543:2000/Cor.1: 2003)

EN ISO 3668: 2001, Paints and varnishes - Visual comparison of the colour of paints (ISO 3668: 1998)

ASTM D 3794: 2013, Standard guide for testing coil coatings

ASTM D 4145-10: 2010, Standard test method for coating flexibility of prepainted sheet

ASTM D 4214-07: 2007, Standard test methods for evaluating the degree of chalking of exterior paint films

EN ISO 4623-1: 2002, Paints and varnishes - Determination of resistance to filiform corrosion - Part 1: Steel substrates (ISO 4623-1: 2000)

EN ISO 4623-2: 2004/AC: 2006, Paints and varnishes - Determination of resistance to filiform corrosion - Part 2: Aluminium substrates (ISO 4623-2: 2003/Cor.1: 2005)

EN ISO 4628-1: 2003, Paints and varnishes - Evaluation of degradation of coatings - Designation of quantity and size of defects, and of intensity of uniform changes in appearance - Part 1: General introduction and designation system (ISO 4628-1: 2003)

EN ISO 4628-2: 2003, Paints and varnishes - Evaluation of degradation of coatings - Designation of quantity and size of defects, and of intensity of uniform changes in appearance - Part 2: Assessment of degree of blistering (ISO 4628-2: 2003)

EN ISO 4628-3: 2003, Paints and varnishes - Evaluation of degradation of coatings - Designation of quantity and size of defects, and of intensity of uniform changes in appearance - Part 3: Assessment of degree of rusting (ISO 4628-3: 2003)

EN ISO 4628-4: 2003, Paints and varnishes - Evaluation of degradation of coatings - Designation of quantity and size of defects, and of intensity of uniform changes in appearance - Part 4: Assessment of degree of cracking (ISO 4628-4: 2003)

EN ISO 4628-5: 2003, Paints and varnishes - Evaluation of degradation of coatings - Designation of quantity and size of defects, and of intensity of uniform changes in appearance - Part 5: Assessment of degree of flaking (ISO 4628-5: 2003)

EN ISO 4628-6: 2011, Paints and varnishes - Evaluation of degradation of coatings - Designation of quantity and size of defects, and of intensity of uniform changes in appearance - Part 6: Assessment of degree of chalking by tape method (ISO 4628-6: 2011)

EN ISO 4628-7: 2003, Paints and varnishes - Evaluation of degradation of coatings - Designation of quantity and size of defects, and of intensity of uniform changes in appearance - Part 7: Assessment of degree of chalking by velvet method (ISO 4628-7: 2003)

EN ISO 4628-8: 2012, Paints and varnishes - Evaluation of degradation of coatings - Designation of quantity and size of defects, and of intensity of uniform changes in appearance - Part 8: Assessment of degree of delamination and corrosion around a scribe or other artificial defect (ISO 4628-8: 2012)

EN ISO 4628-10: 2003, Paints and varnishes - Evaluation of degradation of coatings - Designation of quantity and size of defects, and of intensity of uniform changes in appearance - Part 10: Assessment of degree of filiform corrosion (ISO 4628-10: 2003)

EN ISO 4892-1: 2000, Plastics - Methods of exposure to laboratory light sources - Part 1: General guidance (ISO 4892-1: 1999)

EN ISO 4892-2: 2013, Plastics - Methods of exposure to laboratory light sources - Part 2: Xenon-arc lamps (ISO 4892-2: 2013)

EN ISO 4892-3: 2006, Plastics - Methods of exposure to laboratory light sources - Part 3: Fluorescent UV lamps (ISO 4892-3: 2006)

EN ISO 6270-1: 2001, Paints and varnishes - Determination of resistance to humidity - Part 1: Continuous condensation (ISO 6270-1: 1998)

EN ISO 6270-2: 2005, Paints and varnishes - Determination of resistance to humidity - Part 2: Procedure for exposing test specimens in condensation-water atmospheres (ISO 6270-2: 2005)

EN ISO 6272-1: 2011, Paints and varnishes - Rapid-deformation (impact resistance) tests - Part l: Falling-weight test, large-area indenter (ISO 6272-1: 2011)

EN ISO 6272-2: 2011, Paints and varnishes - Rapid-deformation (impact resistance) tests - Part 2: Falling-weight test, small-area indenter (ISO 6272-2: 2011)

EN ISO 6504-1: 2006, Paints and varnishes - Determination of hiding power - Part 1: Kubelka-Munk method for white and light-coloured paints (ISO 6504-1: 1983)

EN ISO 6504-3: 2007, Paints and varnishes - Determination of hiding power - Part 3: Determination of contrast ratio of light-coloured paints at a fixed spreading rate (ISO 6504-3: 2007)

EN ISO 6721-1: 2011, Plastics - Determination of dynamic mechanical properties - Part 1: General principles (ISO 6721-1:2011)

ISO 6721-11: 2012, Plastics - Determination of dynamic mechanical properties - Part 11: Glass transition temperature [17]

EN ISO 6860: 2006, Paints and varnishes - Bend test (conical mandrel) (ISO 6860: 2006)

EN ISO 7784-1: 2006, Paints and varnishes - Determination of resistance to abrasion - Part 1: Rotating abrasive-paper-covered wheel method (ISO 7784-1: 1997)

EN ISO 7784-2: 2006, Paints and varnishes - Determination of resistance to abrasion - Part 2: Rotating abrasive rubber wheel method (ISO 7784-2: 1997)

EN ISO 7784-3: 2006, Paints and varnishes - Determination of resistance to abrasion - Part 3: Reciprocating test panel method (ISO 7784-3: 2000)

EN ISO 9117-2: 2010, Paints and varnishes - Drying tests - Part 2: Pressure test for stackability (ISO 9117-2: 2010)

EN ISO 9117-6: 2012, Paints and varnishes - Drying tests - Part 6: Print-free test (ISO 9117-6: 2012)

EN ISO 9227: 2012, Corrosion tests in artificial atmospheres - Salt spray tests (ISO 9227: 2012)

EN ISO 11341: 2004, Paints and varnishes - Artificial weathering and exposure to artificial radiation - Exposure to filtered xenon-arc radiation (ISO 11341: 2004)

EN ISO 11507: 2007, Paints and varnishes - Exposure of coatings to artificial weathering - Exposure to fluorescent UV lamps and water (ISO 11507: 2007)

EN ISO 11664-1: 2011, Colorimetry - Part 1: CIE standard colorimetric observers (ISO 11664-1: 2007)

EN ISO 11664-2: 2011, Colorimetry - Part 2: CIE standard illuminants (ISO 11664-2: 2007)

EN ISO 11664-3: 2013, Colorimetry - Part 3: CIE tristimulus values (ISO 11664-3: 2012)

EN ISO 11664-4: 2011, Colorimetry - Part 4: CIE 1976 L*a*b* Colour space (ISO 11664-4: 2008)

EN ISO 11664-5: 2011, Colorimetry - Part 5: CIE 1976 L*u*v* Colour space and u‘, v‘ uniform chromaticity scale diagram (ISO 11664-5: 2009)

EN ISO 11890-1: 2007, Paints and varnishes - Determination of volatile organic compound (VOC) content - Part 1: Difference method (ISO 11890-1: 2007)

EN ISO 11890-2: 2013, Paints and varnishes - Determination of volatile organic compound (VOC) content - Part 2: Gas-chromatographic method (ISO 11890-2: 2013)

EN ISO 11997-1: 2006, Paints and varnishes - Determination of resistance to cyclic corrosion conditions - Part 1: Wet (salt fog)/dry/humidity (ISO 11997-1: 2005) (includes processing acc. VDA 621-415: Cycle B, Appendix D [18])

EN ISO 11997-2: 2006, Paints and varnishes - Determination of resistance to cyclic corrosion conditions - Part 2: Wet (salt fog)/dry/humidity/UV light (ISO 11997-2: 2000)

EN ISO 12137: 2011, Paints and varnishes – Determination of mar resistance (ISO 12137-1: 2011)

EN ISO 13803: 2004, Paints and varnishes – Determination of reflection haze on paint films at 20 degrees (ISO 13803: 2000)

EN 15042-1: 2006, Thickness measurement of coatings and characterization of surfaces with surface waves – Part 1: Guide to the determination of elastic constants, density and thickness of films by laser induced surface acoustic waves

EN 15042-2: 2006, Thickness measurement of coatings and characterization of surfaces with surface waves – Part 2: Thickness measurement of coatings by photothermic method

EN ISO 15110: 2013, Paints and varnishes – Artificial weathering including acidic deposition (ISO 15110: 2013)

EN 15184: 2012, Paints and varnishes – Determination of film hardness by pencil test (ISO 15184: 2012)

EN 15457: 2007, Paints and varnishes – Laboratory method for testing the efficacy of film preservatives in coating against fungi

EN 15458: 2008, Paints and varnishes – Laboratory method for testing the efficacy of film preservatives in coating against algae

EN ISO 15710: 2006, Paints and varnishes – Corrosion testing by alternate immersion in and removal from a buffered sodium chloride solution (ISO 15710: 2002)

EN ISO 15711: 2004, Paints and varnishes – Determination of resistance to cathodic disbonding of coatings exposed to sea water (ISO 15711: 2003)

prEN 16492: (under approval: 2014), Paints and varnishes – Evaluation of the surface disfigurement caused by fungi and algae on coatings

EN ISO 16773-1: 2007, Paints and varnishes – Electrochemical impedance spectroscopy (EIS) in high-impedance coated specimens – Part 1: Terms and definitions (ISO 16773-1: 2007)

EN ISO 16773-2: 2007, Paints and varnishes – Electrochemical impedance spectroscopy (EIS) on high-impedance coated specimens – Part 2: Collection of data (ISO 16773-2: 2007)

EN ISO 16773-3: 2009, Paints and varnishes – Electrochemical impedance spectroscopy (EIS) on high-impedance coated specimens – Part 3: Processing and analysis of data from dummy cells (ISO 16773-3: 2009)

EN ISO 16773-4: 2009, Paints and varnishes – Electrochemical impedance spectroscopy (EIS) on high-impedance coated specimens – Part 4: Examples of spectra of polymer-coated specimens (ISO 16773-4: 2009)

EN ISO 16805: 2005, Binders for paints and varnishes – Determination of glass transition temperature (ISO 16805: 2003)

EN ISO 16862: 2006, Paints and varnishes – Evaluation of sag resistance (ISO 16862: 2003)

EN ISO 17132: 2007, Paints and varnishes – T-bend test (ISO 17132: 2007)

EN ISO 17872: 2007, Paints and varnishes – Guidelines for the introduction of scribe marks through coatings on metallic panels for corrosion testing (ISO 17872: 2007)

EN ISO 20567-1: 2006, Paints and varnishes – Determination of stone-chip resistance of coatings – Part 1: Multi-impact testing (ISO 20567-1: 2005)

EN ISO 20567-2: 2006, Paints and varnishes - Determination of stone-chip resistance of coatings - Part 2: Single-impact test with a guided impact body (ISO 20567-2: 2005)

EN ISO 21227-1: 2003, Paints and varnishes - Evaluation of defects on coated surfaces using optical imaging - Part 1: General guidance (ISO 21227-1: 2003)

EN ISO 21227-2: 2006, Paints and varnishes - Evaluation of defects on coated surfaces using optical imaging - Part 2: Evaluation procedure for multi-impact stone-chipping test (ISO 21227-2: 2006)

EN ISO 21227-3: 2007, Paints and varnishes - Evaluation of defects on coated surfaces using optical imaging - Part 3: Evaluation of delamination and corrosion around a scribe (ISO 21227-3: 2007)

EN ISO 21227-4: 2008, Paints and varnishes - Evaluation of defects on coated surfaces using optical imaging - Part 4: Evaluation of filiform corrosion (ISO 21227-4: 2008)

EN 23270: 1991, Paints and varnishes and their raw materials - Temperatures and humidities for conditioning and testing (ISO 3270: 1984)

10.5.6 Terms and words of art for coatings, coating materials and plastics, and country codes: Standards

EN ISO 1043-1: 2011, Plastics - Symbols and abbreviated terms - Part l: Basic polymers and their special characteristics (ISO 1043-1: 2011)

EN ISO 3166-1: 2006/AC: 2007), Codes for the representation of countries and their subdivisions - Part l: Country codes (ISO 3166-1: 2006/Cor 1: 2007)

EN ISO 4618: 2006, Paints and varnishes - Terms and definitions (ISO 4618: 2006)

EN ISO 18594: 2007, Resistance spot-, projection- and seam-welding - Method for determining the transition resistance on aluminium and steel (ISO 18594: 2007)

10.5.7 Building components: Standards on products and test methods

(in numerical order)

10.5.7.1 CEN bodies in charge

CEN/TC 33, Doors, windows, shutters, building hardware and curtain walling

CEN/TC 127, Fire safety in building

CEN/TC 128, Roof covering, products for discontinuous laying and products for wall cladding

CEN/TC 132, Aluminium and aluminium alloys

CEN/TC 351, Assessment of release of dangerous substances

10.5.7.2 Building components standards

EN 502: 2013, Roofing products from metal sheet - Specification for fully supported roofing products of stainless steel sheet

EN 505: 2013, Roofing products from metal sheet - Specification for fully supported roofing products of steel sheet

EN 507: 1999, Roofing products from metal sheet - Specification for fully supported roofing products of aluminium sheet

EN 508-1: 2008, Roofing products from metal sheet - Specification for self-supported roofing products of steel, aluminium or stainless steel sheet - Part 1: Steel

EN 508-2: 2008, Roofing products from metal sheet - Specification for self-supported roofing products of steel, aluminium or stainless steel sheet - Part 2: Aluminium

EN 508-3: 2008, Roofing products from metal sheet - Specification for self-supported roofing products of steel, aluminium or stainless steel sheet - Part 3: Stainless steel

EN 612: 2005, Eaves gutters with bead stiffened fronts and rainwater pipes with seamed joints made of metal sheet

EN 10162: 2003, Cold rolled steel sections - Technical delivery conditions - Dimensional and cross-sectional tolerances

EN 12219: 1999, Doors - Climatic influences - Requirements and classification

EN 12433-1: 1999, Industrial, commercial and garage doors and gates - Terminology - Part 1: Types of doors

EN 12433-2: 1999, Industrial, commercial and garage doors and gates - Terminology - Part 2: Parts of doors

EN 12519: 2004, Windows and pedestrian doors - Terminology

EN ISO 12944-1: 1998, Paints and varnishes - Corrosion protection of steel structures by protective paint systems - Part l: General introduction (ISO 12944-1: 1998)

EN ISO 12944-2: 1998, Paints and varnishes - Corrosion protection of steel structures by protective paint systems - Part 2: Classification of environments (ISO 12944-2: 1998)

EN 13119: 2007, Curtain walling - Terminology

EN 13830: 2003, Curtain walling - Product standard

EN 14351-1: 2006+A1: 2010, Windows and doors - Product standard, performance characteristics - Part l: Windows and external pedestrian doorsets without resistance to fire and/ or smoke leakage characteristics

EN 14509: 2006/AC: 2008, Self-supporting double skin metal faced insulating panels - Factory made products - Specifications

EN 14782: 2006, Self-supporting metal sheet for roofing, external cladding and internal lining - Product specification and requirements

EN 14783: 2013, Fully supported metal sheet and strip for roofing, external cladding and internal lining - Product specification and requirements

10.5.7.3 Special test standards and features for fire protection

(in numerical order)

EN ISO 1182: 2010, Reaction to fire tests for building products - Non-combustibility test (ISO 1182: 2010)

EN 1363-1: 2012, Fire resistance tests - Part 1: General requirements

EN 1363-2: 1999, Fire resistance tests - Part 2: Alternative and additional procedures

EN 1364-1: 1999, Fire resistance tests for non-loadbearing elements - Part 1: Walls

EN 1364-2: 1999, Fire resistance tests for non-loadbearing elements - Part 2: Ceilings

EN 1364-3: 2006, Fire resistance tests for non-loadbearing elements - Part 3: Curtain walling - Full configuration (complete assembly)

EN 1364-4: 2007, Fire resistance tests for non-loadbearing elements - Part 4: Curtain walling - Part configuration

EN 1365-1: 2012/AC: 2013, Fire resistance tests for loadbearing elements - Part 1: Walls

EN 1365-2: 1999, Fire resistance tests for loadbearing elements - Part 2: Floors and roofs

EN 1365-5: 2004, Fire resistance tests for loadbearing elements - Part 5: Balconies and walkways

EN 13241-1: 2003+A1:2001, Industrial, commercial and garage doors and gates - Product Standard - Part 1: Products without fire resistance or smoke control characteristics

EN 13501-1: 2007+A1: 2009, Fire classification of construction products and building elements - Part l: Classification using data from reaction to fire

EN 13501-5: 2005+A1: 2009, Fire classification of construction products and building elements - Part 5: Classification using data from external fire exposures to roof tests

EN 14351-1: 2006+A1: 2010, Windows and doors - Product standard, performance characteristics - Part l: Windows and external pedestrian doorsets without resistance to fire and/ or smoke leakage characteristics

10.5.8 Quality management and environmental management systems: Standards

EN ISO 9000: 2005, Quality management systems - Fundamentals and vocabulary (ISO 9000: 2005)

EN ISO 9001: 2008/AC: 2009, Quality Management Systems - Requirements (ISO 9001: 2008/Cor. 1: 2009)

EN ISO 9004: 2009, Environmental management systems - Guidelines for management performance improvements (ISO 9004: 2009)

EN ISO 14001: 2004/AC: 2009, Environmental management systems - Requirements with guidance for use (ISO 14001: 2004/Cor 1: 2009)

EN ISO 14004: 2010, Environmental management systems - General guidelines on principles, systems and support techniques (ISO 14004: 2004)

EN ISO 14015: 2010, Environmental management systems - Environmental assessment of sites and organizations (EASO) (ISO 14015: 2001)

EN ISO 14020: 2001, Environmental labels and declarations - General principles (ISO 14020: 2000)

EN ISO 14021: 2001/A1: 2011, Environmental labels and declarations - Self-declared environmental claims (Type II environmental labelling) (ISO 14021: 1999/Amd 1: 2011)

EN ISO 14024: 2000, Environmental labels and declarations - Type I environmental labelling - Principles and procedures (ISO 14024: 1999)

EN ISO 14031: 1999, Environmental management - Environmental performance evaluation - Guidelines (ISO 14031: 1999)

EN ISO 14040: 2006, Environmental management - Life cycle assessment - Principles and framework (ISO 14040: 2006)

EN ISO 14044: 2006, Environmental management - Life cycle assessment - Requirements and guidelines (ISO 14044: 2006)

EN ISO 19011: 2011, Guidelines for quality and/or environmental management systems auditing (ISO 19011: 2011)

10.5.9 Selected European organisations

APEAL - Association of European Producers of Steel for Packaging; Brussels, www.apeal.org

CEPE - The European Council of the Paint, Printing Ink and Artists' Colours Industry/Conseil Européen de l'Industrie des Peintures, des Encres d'Imprimerie et des Couleurs d'Art, EU Sectors Can Coatings, Coil Coatings, Powder Coatings, Clean Air For Europe (CAFE), The European Chemical Policy (REACH), The Globally Harmonised Svstem of Classification and Labelling of Chemicals (GHS), Brussels, www.cepe.org

Construction Products Europe (formerly: CEPMC), Brussels, www.construction-products.eu

EAA - European Aluminium Association, Brussels, www.alueurope.eu

ECCA - European Coil Coating Association, Brussels, www.prepaintedmetal.eu

Eurofer - European Federation of Iron and Steel Industries, Brussels, www.eurofer.org

EUROMETAL, European Federation of Steel Distributors, Brussels, www.eurometal.eu

IAI - International Aluminium Institute, London, www.world-aluminium.org

IISI - Intemational Iron and Steel Institute, Brussels, www.worldsteel.org

RTE - RadTech Europe, Den Haag, www.radtech-europe.com

10.6 Literature

[1] Meuthen, B., Jandel, A.-S., Coil Coating, 2nd ed., Vieweg, Wiesbaden 2008, pp. 307 ff

[2] Meuthen, B., Jandel, A.-S [1], pp. 316 ff

[3] esearch.cen.eu/esearch/extendedsearch.aspx (29-06-13; 13:21 h)

[4] www.astm.org/Standard/index.shtml (29-06- 13; 13:32 h)

[5] eur-lex.europa.eu/... (29-06-13; 13:49 h)

[6] ftp.cencenelec.eu/CEN/Products/Latestpublications/LatestPublications_2013_May.pdf (29-06-13; 14:01 h)

[7] www.cen.eu/cen/Members/Pages/default.aspx (12-03-13, 14:31 h)

[8] www.iso.org/iso/home/about.htm

[9] www.astm.org/ABOUT/aboutASTM.html (12-03-13, 19:06 h)

[10] anon., SEW 022: Kontinuierlich schmelztauchveredelte Flacherzeugnisse aus Stahl - Zink-Magnesium-Überzüge, techn. Lieferbedingungen (Continuously hot-dip coated steel flat products - zinc-magnesium coatings, techn. delivery conditions) (08.10), Stahleisen, Düsseldorf 2012

[11] www.cenelec.eu/dyn/www/f?p=104:110:2408623675329485::::FSP_PROJECT,FSP_LANG_ID: 21933,25; 11-09-13, 10:41 h

[12] www.cenelec.eu/dyn/www/f?p=104:11:3879827928076213::::FSP_PROJECT,FSP_LANG_ID: 23419,25; 10-09-13, 13:27 h

[13] www.cenelec.eu/dyn/www/f?p=104:110:4441611114693449::::FSP_PROJECT: 22542,25; 05-11-13, 11:07 h

[14] Alanen, A., private communication, European Coil Coating Association, Brussels 2013

[15] anon., Standard test method for assignation of the glass transition temperature by dynamic mechanical analysis, ASTM, W. Conshohocken, PA, USA, 2009; www.astm.org/Standards/E1640.htm (27-07-13; 11:25 h)

[16] anon., Cyclic corrosion testing of materials and components in automotive construction, Stahl-Eisen Prüfblatt (SEP) 1850, 1st ed., (equivalent to VDA 233-102), Stahleisen eds., Düsseldorf 2012

[17] anon., Plastics - Determination of dynamic mechanical properties - Part 11: Glass transition temperature, ISO, Geneva 2012; www.iso.org/iso/catalogue_detail.htm?csnumber=52845 (27-07-13; 11:34 h)

[18] anon., DIN EN ISO 11997-1: 2006-04, Paints and varnishes - Determination of resistance to cyclic corrosion conditions - Part 1: Wet (salt fog)/dry/humidity (ISO 11997-1: 2005), German version: Beuth, Düsseldorf 2006, www.beuth.de/de/norm/din-en-iso-11997-1/87717570 (27-06-13; 21:08 h)

Author

Dr. Jörg Sander, born 1958, obtained his Doctor's degree in chemistry at the University of Münster, Germany. In 1987 he joined the Henkel Group, where he dealt with process chemicals for the surface technology of metals from both a technical and a commercial perspective for more than 20 years. Serving the target markets of steel and aluminium strip making and finishing, the production of beverage cans and the wire and extrusion industry, he managed teams in the research, product development, sales, and marketing functions. Between 1996 and 2009 he was Official Delegate for the Henkel Group to the European Coil Coating Association (ECCA), where he chaired the Appliance Task Group and related marketing project teams. Currently, he is working as self-employed industry consultant and free scientific writer.

Index

Symbols

A

B

C

D

E

O

P

Q

R

S

T